Lecture Notes in Mathematics

Edited by A. Dold and B. Eckmann

565

Turbulence and Navier Stokes Equations

Proceedings of the Conference
Held at the University of Paris-Sud
Orsay June 12-13 1975

Edited by Roger Temam

Springer-Verlag
Berlin · Heidelberg · New York 1976

Editor
Roger Temam
Mathématique
Bâtiment 425
Université de Paris-Sud
Centre d'Orsay
91405 Orsay/France

Library of Congress Cataloging in Publication Data

Main entry under title:

Turbulence and Navier Stokes equations.

(Lecture notes in mathematics ; 565)
English and French.
1. Turbulence--Congresses. 2. Navier-Stokes equations--Congresses. I. Temam, Roger. II. Series: Lecture notes in mathematics (Berlin) ; 565.
QA3.L28 no. 565 [QA913] 510'.8s [532'.0527'01515352]
76-51758

AMS Subject Classifications (1970): 76D05, 76F05, 60H15, 35K45, 65P05, 35R10

ISBN 3-540-08060-0 Springer-Verlag Berlin · Heidelberg · New York
ISBN 0-387-08060-0 Springer-Verlag New York · Heidelberg · Berlin

PREFACE

On a pu noter ces dernières années un important regain d'intérêt pour les aspects mathématiques de la turbulence :

- Travaux de topologie et de géométrie diffétentielle sur les propriétés asymptotiques des solutions d'équations d'évolution dissipatives (attracteurs, ...).

- Nouveaux résultats d'existence et de régularité pendant un temps fini pour les équations d'Euler en dimension 3.

- Travaux sur les équations de Navier Stokes aléatoires traitées soit directement soit à l'aide de mesures sur des espaces fonctionnels.

- Mise en forme mathématique des modèles stochastiques de la turbulence conduisant dans la limite de viscosité nulle à des singularités au bout d'un temps fini et à des spectres en loi de puissance.

- Introduction d'objets géométriques de dimension de Hausdorff fractionnaire pour représenter la partie dissipative d'un écoulement turbulent dans la limite de viscosité nulle.

- Mise au point de techniques de simulation numérique directe des équations de Navier Stokes aléatoires.

Pour permettre de faire le point sur ces recherches, la Société Mathématique de France a proposé l'organisation de Journées Mathématiques sur la Turbulence. Celles-ci ont eu lieu à l'Université de Paris-Sud à Orsay, les 12 et 13 Juin 1975. Ces Journées étaient également patronées par le Centre National de la Recherche Scientifique dans le cadre de l'Action Thématique programmée "Instabilité et Turbulence" et parle Comité Francais des Mathématiciens.

Les Journées ont réuni plus d'une centaine de Mathématiciens, mécaniciens, physiciens et numériciens, français et étrangers, intéressés par les aspects mathématiques de la turbulence. Elles ont permis des contacts interdisciplinaires très stimulants dont ces actes ne rendent évidemment pas compte. Nous reproduisons ici le texte des Conférences présentées.

Je remercie les membres du Comité scientifique pour leur participation à l'organisation de ces journées. Les présidents de séance et tous les participants ont beaucoup contribué à la réussite de ces Journées par la qualité des discussions suscitées. Qu'ils en soient vivement remerciés.

Enfin je remercie tout particulièrement Mme Maynard qui s'est occupée avec beaucoup de soin du secrétariat de la Conférence.

R. Temam

PREFACE

There has been a strong renewed interest in the mathematical aspects of Turbulence during recent years :

- Topological and differential geometrical approaches to the asymptotic properties of solutions of dissipative equations of evolution (attractors, ...)

- New existence and regularity results during a finite time for the three dimensional Euler equations.

- Direct and function-space approaches to the random Navier-Stokes equations.

- Mathematical investigation of stochastic models in Turbulence leading at vanishing viscosity, to singularities after a finite time and to power law spectra.

- Introduction of geometric objects with fractional Hausdorff dimension to represent the dissipative part of a turbulent flow at vanishing viscosity.

- Parallel to this, direct numerical simulation techniques for the random Navier-Stokes equations have been developed.

In order to examine the present state of the subject the French Mathematical Society has proposed the organisation of a Workshop on the Mathematical aspects of Turbulence. This workshop was held at the University of Paris-South at Orsay on June 12 and 13, 1975. This workshop was also sponsered by the Centre National de la Recherche Scientifique as part of the A.T.P. programm "Instabilité et Turbulence", and by the Comité Français des Mathématiciens.

This Conference has brought together more than a hundred of french or foreign mathematicians, fluid dynamicists, physicists and numerical analysts interested in the mathematical aspects of Turbulence. This was the opportunity of very stimulating interdisciplinary contacts which cannot be reproduced here. We reproduce in these proceedings the text of the lectures presented.

I wish to thank the members of the scientific committee for their participation to the organisation of this workshop. We are very grateful to the session chairmen and all the participants who greatly contributed to the success of the conference by the discussions they raised.

Finally, I whish to thank Mrs Maynard who took care of the secretary of the Conference.

R. Temam

PARTICIPANTS

Comité Scientifique
Scientific Committee

C. BARDOS - Dép. de Math., Fac. des Sci., Parc Valrose, Nice.

M. CRAYA - Dép. de Math., Université de Grenoble, Grenoble.

U. FRISCH - Observatoire de Nice, Nice.

J.L. LIONS - Collège de France, Paris.

D. RUELLE - I.H.E.S., route de Chartres, Bures-sur-Yvette.

R. TEMAM - Dép. de Math., Univ. Paris XI, Centre d'Orsay, Orsay.

Présidents de Séance
Sessions Chairmen

P. GERMAIN - Université Paris VI, place Jussieu, Paris.

G. GUIRAUD - U.E.R. de Mécanique, Univ. Paris VI, Paris.

J. LERAY - Collège de France, Paris.

J.L. LIONS - Collège de France, Paris.

Conférenciers
Lecturers

C. BARDOS - Dép. de Math., Fac. des Sci., Parc Valrose, Nice.

D. G. EBIN - Dep. of Math., Suny at Stony Brook, N.Y., U.S.A.

C. FOIAS - Univ. de Bucarest, Bucarest, Roumanie.

U. FIRSCH - Observatoire de Nice, Nice.

G. IOOSS - I.M.S.P. Université de Nice, Nice

D. D. JOSEPH - Dep. of Math., Univ. of Minnesota, Minneapolis, U.S.A.

J.P. KAHANE - Dép. de Math., Univ. Paris XI, Centre d'Orsay, Orsay.

T. KATO - Dep. of Math., Univ. of California, Berkeley, U.S.A.

O. A. LANFORD - I.H.E.S., route de Chartres, Bures-sur-Yvette.
et Dep. of Math., Univ. of California, Berkeley, U.S.A.

M. LESIEUR - Observatoire de Nice, Nice.

B. MANDELBROT - I.B.M. Reseach Center, Yorktown Heights, N.Y., U.S.A.

J. E. MARSDEN - Dep. of Math., Univ. of California, Berkeley, U.S.A.

S. ORSZAG - Mass. Inst. of Tech., Cambridge, Mass. U.S.A.

P. PENEL - Dép. of Math., Univ. Paris XI, Centre d'Orsay, Orsay.

D. RUELLE - I.H.E.S., route de Chartres, Bures-sur-Yvette.

D. H. SATTINGER - Univ. of Minnesota, Minneapolis, Minn. U.S.A.

V. SCHEFFER - Observatoire de Nice, Nice.

P.L. SULEM - Observatoire de Nice, Nice.

R. TEMAM - Dép. de Math., Univ. Paris XI, Centre d'Orsay, Orsay.

TABLE DES MATIÈRES

CONTENTS

FINITE-TIME REGULARITY FOR BOUNDED AND UNBOUNDED IDEAL INCOMPRESSIBLE FLUIDS USING HOLDER ESTIMATES

C. BARDOS * and U. FRISCH **

* *Département de Mathématiques, Université de Nice, Parc Valrose, 06034 NICE*

** *Observatoire de Nice, B.P. 252, 06007 NICE Cedex.*

I INTRODUCTION

We are concerned in this paper with the existence, uniqueness and regularity during a finite time of solutions of the Euler equation for an ideal three-dimensional in-compressible fluid

(1) $\frac{\partial u}{\partial t} + u.\nabla u = -\nabla p \;;\; \nabla.u = 0,$

plus initial conditions u_0 and boundary conditions on a domain Ω.

It is known that sufficiently high smoothness of initial data will persist at least to some (data-dependent) time T* .The first such result goes back to L.LICHTENSTEIN [11] where it is assumed that the initial vorticity, the curl of u_0, has compact support and where persistence of regularity (namely Hölder continuity of the vorticity) is shown only up to a time such that the fractional change in the vorticity is small compared to unity. More recently D.EBIN and J.MARSDEN [5] analyzed the case of flows in a bounded domain, using the fact that the solution is a geodesic flow on an infinite dimensional manifold; they were able to prove persistence, during a finite time of C^∞ regularity in space. Similar results can also be obtained by simpler Sobolev-space techniques which, in addition yield explicit expressions of T* (T.KATO [7] R.TEMAM [12] ; C.FOIAS, U.FRISCH and R.TEMAM [6]) . In all these results T* depends on a suitable Sobolev norm of u_0 in such a way that T* goes to zero when the total energy of u_0 goes to infinity. This rules out applications to flows in unbounded domains with no decay at infinity and, in particular, to homogenous turbulence which has a finite energy per unit volume but infinite total energy (S. ORSZAG [3], M. LESIEUR and P.L.SULEM [10]) . Actually, it is unlikely that the best possible choice of T* should depend on the total energy; indeed, there are several reasons to believe that T* should be of the order of the inverse of the maximum of the modulus of the initial vorticity (or velocity gradient) : First, for the Burgers equation (with viscosity dropped), one can show that the first singularity appears at $T^* = [\sup - \frac{du_0}{dx}]^{-1}$ Second, there is a "vortex-streching" heuristic argument : we take the curl of (1) to obtain (with $\omega = \nabla \wedge u$ and $\frac{D}{Dt}$ denoting the Lagrangian derivative)

$$\frac{D\omega}{Dt} = \frac{\partial \omega}{\partial t} + u.\nabla \omega = \omega.\nabla u \quad .$$

Notice that the vorticity is just the antisymmetric part of the velocity gradient ∇u; if we tentatively identify ω and ∇u and discard vector and tensor indices we obtain $\frac{D\omega}{Dt} = \omega^2$ which implies that ω blows up at the time $T^* = |\sup \omega_0|^{-1}$. This argument suggest trying to work with a sup-norm rather than a Sobolev-norm e.g. $\sup|\omega|$ (Norms involving space integral may be called "extensive" and those involving sup operations "intensive". Intensive norms seem more appropriate for fluid dynamics because what happens to a given fluid element is only very weakly affected by distant fluid elements (through the pressure term)). In fact the sup norm of ω is not quite sufficient because the boundedness of ω does not imply that of ∇u; however, such boundedness is ensured if we take ω Hölder-continuous.

We shall actually show in this paper that if u_0 and ω_0 are Hölder continuous they remain so at least to some T^* independent of the total energy which may even be infinite. The proof holds both for bounded and unbounded domains. C^∞ regularity (assumed initially) is then easily proved (by taking derivatives of the Euler equation) on the interval $[0,T^*[$.

An open problem is the analyticity in $[0,T^*[$ although some recent very general results of S. BAOUENDI and C. GOULAOUIC [2] suggests its persistence. A much more difficult problem is to show that singularities do actually occur in the three dimensional Euler equation. So far this has been shown only on the so called "spectral equations" of homogenous isotropic turbulence which are not exact consequences of the Euler equation [1,3,10]. Finally the study of spectral equations [1] suggests that, in addition to the well known energy estimate, there may exist certain uniform estimates in $[0, T^*[$; a precisely formulated conjecture would be premature but the theoretical and experimental results on the Kolmogorov-spectrum (the $k^{-5/3}$ law [3]) suggest that the spatial Fourier transform of u should decrease at infinity faster than some inverse power of the modulus of the wave-vector.

II NOTATIONS AND STATEMENT OF THE RESULTS

Ω denotes an open set of R^n bounded or unbounded ; we shall assume that the boundary of Ω, $\partial\Omega$ is a smooth compact manifold ; for any $x \in \partial\Omega$,

$$\nu(x) = (\nu_1(x), \nu_2(x), \ldots, \nu_n(x))$$

will denote the outward normal to $\partial\Omega$; generally n will be taken equal to 2 or 3. We shall denote by $C^{0,\alpha}(\Omega)$ (in short $C^{0,\alpha}$) the space of vector valued functions such that

$$(2) \qquad \sup_{x \in \Omega} |u(x)| + \sup_{\substack{x,y \in \Omega \times \Omega \\ x \neq y}} \left(|u(x) - u(y)| / |x-y|^{\alpha} \right) < +\infty$$

$|u|$ denotes the Euclidian norm of the vector u (or the tensor h) , the left hand side of (2) will be denoted by $\|u\|_{0,\alpha}$.

Similarly we shall denote by $C^0(\Omega)$ the space of continuous bounded functions defined on Ω and we shall write :

$$\|u\|_0 = \sup_{x \in \Omega} |u(x)|$$

Finally we shall denote by $C^{k,\alpha}(\Omega)$ the following space .

$$C^{k,\alpha}(\Omega) = \{u \in C^0 \, , \, D^{\ell}u \in C^0 \, , \, |\ell| < k \, ; \, D^{\ell}u \in C^{\alpha} \, , \, |\ell| = k\}$$

equipped with the norm

$$\|u\|_{k,\alpha} = \sum_{|\ell| < k} \|D^{\ell}u\|_0 + \sum_{|\ell| = k} \|D^{\ell}u\|_{0,\alpha}$$

All these spaces are Banach spaces, and it is convenient to recall that for any continuously differentiable function $t \to u(t)$ with value in a Banach space B one has

$$(3) \qquad \frac{d}{dt_+} \|u\| \leq \| \frac{du}{dt} \|$$

where $\frac{d}{dt_+}$ denotes the right derivative ; $\nabla.,\nabla,\nabla\wedge$ and $u.\nabla$ denote the usual operators divergence, gradient, curl and $\sum_{i=1}^{n} u_i \frac{\partial u}{\partial x_i}$, the sign Σ will be generally omitted . We shall denote by $C^{k,\alpha}_{\sigma}$ the spaces :

$$C^{k,\alpha}_{\sigma} = \{u \in C^{k,\alpha}(\Omega) \, , \, \nabla.u = 0 \, , \, \nu(.).u(.)|_{\partial\Omega} = 0\} \quad .$$

Since the differential system $\nabla . u = f$, $\nabla \wedge u = \omega$, $u(.)|_{\partial\Omega} = 0$ is elliptic there exists some constants $c_\alpha^k(\Omega)$ such that the following estimates hold for any $u \in C_\sigma^{k,\alpha}(\Omega)$:

$$(4) \quad \begin{cases} \|u\|_{k,\alpha} \leq c_\alpha^k(\Omega) \, \|\nabla \wedge u\|_{k-1,\alpha} & \text{when } \Omega \text{ is bounded} \\ \|u\|_{k,\alpha} \leq c_\alpha^k(\Omega)(\|\nabla \wedge u\|_{k-1,\alpha} + \|u\|_{k-1,\alpha}) & \text{when } \Omega \text{ is unbounded} \end{cases}$$

Therefore

$$|||u|||_{k,\alpha} = \begin{cases} \|\nabla \wedge u\|_{k-1,\alpha} & \text{when } \Omega \text{ is bounded} \\ \|\nabla \wedge u\|_{k-1,\alpha} + \|u\|_{k-1,\alpha} & \text{when } \Omega \text{ is unbounded} \end{cases}$$

Will define on $C_\sigma^{k,\alpha}(\Omega)$ an equivalent norm.

Now we state the results.

THEOREM 1 : Let Ω be a bounded set of $\mathbb{R}^3$; then there exists a constant $C(\Omega,\alpha) = C_1$ depending on Ω and α $(0 < \alpha < 1)$ such that for every $u_0 \in C_\sigma^{1,\alpha}$ there exists a unique function $u \in C(-T^*,T^* ; C_\sigma^{1,\alpha})$ which is a solution of the Euler equation and which, for $t = 0$ is equal to u_0. Furthermore T^* is given by the relation $T^* = (C_1 \| u_0 \|_{1,\alpha})^{-1}$, and u satisfies the a priori estimate

$$(5) \quad |||u(t)|||_{1,\alpha} \leq |||u_0|||_{1,\alpha} \frac{T^*}{T^* - |t|} .$$

When Ω is an unbounded open set of $\mathbb{R}^3$ containing the exterior of a ball (Ω possibly equal $\mathbb{R}^3$) we have

THEOREM 2 : There exists a constant C_2 and a bilinear continuous map $(u,v) \to F(u,v)$ defined on $C^{1,\alpha} \times C^{0,\alpha}$ with value in $C^{0,\alpha}$ with the following properties.

1. $F(u,u)$ is a gradient
2. For every $u_0 \in C_\sigma^{1,\alpha}$ there exists a unique function $u \in C(-T^*, T^* ; C_\sigma^{1,\alpha})$

which is a solution of the problem

$$(6) \qquad \frac{\partial u}{\partial t} + u.\nabla u = F(u,u) , \; u(x,o) = u_0(x) .$$

T^* is given by $T^* = (C_2 |||u_0|||_{1,\alpha})^{-1}$ and the solution satisfies the estimate

$$(7) \qquad |||u(t)|||_{1,\alpha} \leq |||u_0|||_{1,\alpha} \frac{T^*}{T^* - |t|} .$$

3. When $u_o \in C^{1,\alpha}_\sigma$ belongs to L^2 (finite energy) the solution of (6) coincides with the usual solution of the Euler equation

$$\frac{\partial u}{\partial t} + u.\nabla u = -\nabla p \quad (\nabla p \in L^2) , \quad u(x,o) = u_o(x) \quad .$$

PROPOSITION 1. Assume that in the situation of theorem 1 or 2 the initial data belongs to $C^{k,\alpha}_\sigma$ then the solution of the Euler equation (or in theorem 2 the solution of (6)) belongs to $C(-T^*, T^*, C^{k,\alpha}_\sigma)$. In particular if $u_o \in C^\infty$, then $u(t)$ remains in C^∞ on the whole interval $]-T^*, T^*[$.

To state the next result and also for use in the proofs of the theorem 1 and 2 we introduce Lagrangian coordinates. Let $v \in C(-T,T \; ; \; C^{1,\alpha}(\Omega))$ satisfying the boundary condition $v.\nu|_{\partial\Omega x]-T,T[} = 0$, then for any $x \in \Omega$ there exists a unique continuously differentiable mapping $t \to x_v(t)$ (denoted $x(t)$) defined on $|t| < T$ with value in Ω such that one has

$$(7) \qquad x'(t) = v(x(t),t) \; , \; x(o) = x \quad .$$

This is proved for instance in J.P. BOURGUIGNON & H. BREZIS [4] , lemma A.6. The hypothesis $v.\nu|_{\partial\Omega x]-T,T[} = 0$ is essential : Furthermore, for fixed t , the mapping $x \to x(t)$ is a bijection of Ω onto Ω which is measure preserving (Liouville Theorem). For $(x,y) \in \Omega \times \Omega$ we will put $\rho_v(x,y,t) = |x(t)-y(t)|$ (denoted $\rho_v(t)$) .

PROPOSITION 2. Assume that Ω is bounded let $L = \text{Sup} \, |x-y| \; ((x,y) \in \Omega \times \Omega)$. Let $u \in C(-T^*,T^* \; ; \; C^{1,\alpha}_\sigma(\Omega))$ be a solution of the Euler equation then one has

$$(8) \quad L \left| \frac{\rho_u(o)}{L} \right|^{e^{D(\Omega)\int_o^t \|(\nabla\wedge u)(s)\|_o ds}} \leq \rho_u(t) \leq L \left| \frac{\rho_u(o)}{L} \right|^{e^{-D(\Omega)\int_o^t \|(\nabla\wedge u)(s)\|_o ds}}$$

where $D(\Omega)$ denotes a constant depending on Ω .

Remarks 1. Theorems 1, 2 and Proposition 1 are valid in arbitrary dimensions and in particular in two dimensions. However in two dimensions when Ω is bounded global regularity holds (see [2] for the proof where the two dimensional analog of Proposition 2 is used). Actually the result of Proposition 2 is interesting by itself because it relates vorticity and pair dispersion.

2. When Ω is unbounded we dont know if global regularity holds in two dimensions.

3. When u does not decay for $|x| \to \infty$ the relation $\nabla.(u\nabla u) = -\nabla p$ (see lemma 2 , below) determines ∇p up to a constant vector; this is the

reason why we need the operator $F(\cdot,\cdot)$.

III PROOFS OF THE THEOREMS

We shall use the following

LEMMA 1 : Let $u(\cdot,\cdot)$, $v(\cdot,\cdot)$ and $g(\cdot,\cdot)$ be three smooth functions defined on $]-T,T[\times\Omega$ $(T>0)$ satisfying the equations

$$(9)\qquad \frac{\partial u}{\partial t} + v\cdot\nabla u = g \quad , \quad v\cdot\nu|_{\partial\Omega\times]-T,T[} = 0$$

then one has

$$(10)\qquad \frac{d}{dt_+}\|u\|_{0,\alpha} \le \|g\|_{0,\alpha} + \alpha\,\|\nabla v\|_0\|u\|_{0,\alpha}$$

PROOF. We use the mapping $t \to x_v(t)$ defined above, the equation (9) is then written :

$$(11)\qquad \frac{d}{dt}\,u(x(t),t) = g(x(t),t)$$

From (11) we deduce that we have

$$(12)\qquad \frac{d}{dt_+}\|u\|_0 \le \|g\|_0 \quad .$$

Next we write

$$(13)\qquad \|u(t)\|_\alpha = \operatorname*{Sup}_{(\xi,\eta)\in\Omega\times\Omega} \frac{|u(\xi,t)-u(\eta,t)|}{|\xi-\eta|^\alpha}$$

Since $x \to x(t)$ is a bijection we have also

$$(14)\qquad \|u(t)\|_\alpha = \operatorname*{Sup}_{(x,y)\in\Omega\times\Omega} \frac{|u(x(t),t)-u(y(t),t)|}{\rho_v^\alpha(x,y,t)}$$

and from (14) we deduce the inequality :

$$(15)\qquad \frac{d}{dt_+}(\|u(t)\|_\alpha) \le \operatorname*{Sup}_{(x,y)\in\Omega\times\Omega}\left\{\frac{|u'(x(t),t)-u'(y(t),t)|}{\rho(t)^\alpha} + \alpha\frac{|u(x(t),t)-u(y(t),t)|}{\rho(t)^{\alpha+1}}\,\rho'(t)\right\}$$

Since $x'_v(t) - y'_v(t) = v(x(t),t) - v(y(t),t)$, we have

$$(16) \qquad \rho'(t) \le |x'_v(t) - y'_v(t)| \le \|\nabla v\|_0 \, \rho_v(x,y,t)$$

And from (15) and (16) we deduce (10) .

Now we can give the proof of theorem 1 . We introduce the sequence $u^n \in C(\mathbb{R} ; C^{1,\alpha}_\sigma(\Omega))$ defined by the relations

$$u^0(x,t) = u_0(x) \quad , \quad \omega^0(x,t) = \nabla \wedge u_0(x) \quad .$$

$$(17) \quad \begin{cases} \dfrac{\partial \omega^{n+1}}{\partial t} + u^n . \nabla \, \omega^{n+1} = \omega^{n+1} . \nabla \, u^n \quad , \quad \omega^{n+1}(0) = \nabla \wedge u_0 \\[2ex] \nabla . u^{n+1} = 0 \quad , \quad \nabla \wedge u^{n+1} = \omega^{n+1} \quad , \quad u^{n+1} . \nu|_{\partial\Omega \times R} = 0 \end{cases}$$

Now using (10) and (3) we readily obtain :

$$(18) \qquad \frac{d}{dt_+} |||u^{n+1}|||_{1,\alpha} \le C_1 \, |||u^n|||_{1,\alpha} \, |||u^{n+1}|||_{1,\alpha}$$

(for $t > 0$, the case $t < 0$ can be done similarly) .

From (18) we deduce by induction that we have $|||u^{n+1}(t)|||_{1,\alpha} \le y(t)$ where $y(t)$ is the solution of the non linear differential equation

$$(19) \qquad y' = C\, y^2 \, , \quad y(0) = |||u(0)|||_{1,\alpha}$$

From (19) we deduce that on the interval $[0,T^*[$ (and similarly on $]-T^*,0]$)

$$(20) \qquad |||u^{n+1}(t)|||_{1,\alpha} \le \frac{T^*}{T^* - |t|} \, |||u_0|||_{1,\alpha}$$

We denote by $\mathcal{M}$ the set $\{u^0, u^1, u^2, \dots, u^n, \dots\}$ restricted to $]-T^*+\varepsilon, T^*-\varepsilon[$, $\mathcal{M}$ is bounded in $C(]-T^*+\varepsilon \, , \, T^*-\varepsilon \, , \, C^{1,\alpha}_\sigma(\Omega))$ $(\varepsilon > 0)$; we denote by Φ the mapping $u^n \mapsto u^{n+1}$.

From the relations :

$$\frac{\partial \omega^{n+1}}{\partial t} + u^n . \nabla \, \omega^{n+1} = \omega^{n+1} . \nabla \, u^n \quad \text{and} \quad \frac{\partial \omega^{p+1}}{\partial t} + u^p . \nabla \, \omega^{p+1} = \omega^{p+1} . \nabla \, u^p$$

One deduces, using once again lemma 1 and the estimate (3), the inequality :

$$(21) \quad \frac{d}{dt_+} |||u^{n+1} - u^{p+1}|||_{1,\alpha} \le C_3 \, |||u^{n+1} - u^{p+1}|||_{1,\alpha} + C_4 |||u^n - u^p|||_{1,\alpha} \; .$$

(C_3 and C_4 are uniformly bounded for $|t| \leq T^* - \epsilon$) .

From (21) one deduces easily the inequality

$$(22) \quad |||u^{n+m}(t) - u^{p+m}(t)|||_{1,\alpha} \leq (C_5 e^{CT} t)^m/m! \; |||u^n - u^p|||_{1,\alpha}$$

Therefore for m large enough Φ^m is a strict contraction of $\mathcal{M}$ in $\mathcal{M}$. Therefore one can use a fixed point theorem to prove that there exists u such that: $u \in C(-T^* + \epsilon \, , \, T^* - \epsilon \; ; \; C^{1,\alpha}_\sigma)$ solution of the equations

$$(23) \quad \nabla \Lambda u = \omega \, , \; \frac{\partial \omega}{\partial t} + u.\nabla \omega = \omega.\nabla u \; . \; u(x,0) = u_0(x) \; .$$

From the relations (23) one deduces that $\nabla \Lambda (\frac{\partial u}{\partial t} + u \nabla u) = 0$ and therefore, that $\frac{\partial u}{\partial t} + u.\nabla u = - \nabla p$ (c.f. M. ZERNER [13]) . This complete the proof of the existence of the solution for theorem 1 ; (20) gives the estimate (5) when n goes to infinity. The uniqueness is easy and left to the reader .

To consider the case of an unbounded domain we will need

LEMMA 2 : Assume that Ω is an open set of $\mathbb{R}^3$ with smooth boundary, containing the exterior of a ball then there exists a bilinear continuous maps $(u,v) \to F(u,v)$ defined on $C^\alpha(\Omega) \times C^{1,\alpha}(\Omega)$ with value in $C^{0,\alpha}(\Omega)$ with the following properties

(i) For any pair $(u,v) \in C^{1,\alpha} \times C^\alpha$ $F(u,v)$ is a gradient .

(ii) For any pair $(u,v) \in C^{1,\alpha}_\sigma \times C^{1,\alpha}_\sigma$ one has $\nabla.(v.\nabla u - F(u,v)) = 0$.

(iii) If $u \in C(-T^*, T^* \; ; \; C^{1,\alpha} \cap L^2)$ is a finite energy solution of the Euler equation $\frac{\partial u}{\partial t} + u \nabla u = - \nabla p$, $(\nabla p \in L^2)$ one has $- \nabla p = F(u,u)$.

PROOF : For the sake of simplicity and to emphasize the importance of the behaviour at infinity of u , we will give the proof of this lemma only in the case $\Omega = \mathbb{R}^3$, when $\partial\Omega \neq \{\emptyset\}$ the proof is similar, but relies on the analysis of the Green function of the exterior Neumann problem. Taking the divergence of both sides of the Euler equation one obtains :

$$(24) \qquad - \nabla p = \frac{\partial u_i}{\partial x_j} \frac{\partial u_j}{\partial x_i} = \frac{\partial^2}{\partial x_i \partial x_j} (u_i u_j)$$

And if the right hand side of (24) is bounded (in $L^1(\mathbb{R}^n)$ for instance) the only solution (up to a constant) of (24) is given by :

$$(25) \qquad p = K(.) * \frac{\partial^2}{\partial x_i \partial x_j} (u_i u_j) \, , \quad K(x) = \frac{1}{4\pi} |x|^{-1} \; .$$

Now we introduce a smooth function $\theta \in \mathcal{D}(R^3)$ $\theta \equiv 1$ in a neighbourhood of zero, we put $\tilde{\theta} = 1 - \theta$ and write $K(.) = K_1(.) + K_2(.)$ $(K_1 = \theta K$, $K_2 = \tilde{\theta} K$ and we put :

$$(26) \qquad F(u,v) = \left(\frac{\partial}{\partial x_i} (\nabla K_1(.))\right) * v_j \frac{\partial u_i}{\partial x_j} + \left(\frac{\partial^2}{\partial x_i \partial x_j} (\nabla K_2(.))\right) * v_j u_i \quad .$$

$\frac{\partial}{\partial x_i} (\nabla K_1(.))$ is a function with compact support, smooth outside 0 and which behaves near 0 like $|x|^{-3}$, therefore $g \to \left(\frac{\partial}{\partial x_i} (\nabla K_1(.))\right) * g$ is a linear continuous mapping from $C^{0,\alpha}$ to $C^{0,\alpha}$ (this can be proved exactly like in LADYZENSKAIA & URALSTEVA [9] chap.3 § 2) . On the other hand $\frac{\partial^2}{\partial x_i \partial x_j} \nabla K_2(.)$ is a smooth function, its support is not compact but it behaves at infinity like $|x|^{-4}$ therefore it belongs to $L^1(R^3)$ and the mapping $g \to \left(\frac{\partial^2}{\partial x_i \partial x_j} \nabla K_2(.)\right) * g$ is continuous from $C^{0,\alpha}$ to $C^{0,\alpha}$.

We have thus proved that the right hand side of (26) defines a bilinear continuous operator from $C^{1,\alpha}_\sigma \times C^{0,\alpha}$ with value in $C^{0,\alpha}$. It is easy to prove that F does not depend on the choice of the truncation function θ and satisfies the properties (i) and (ii) ; to prove (iii) one notices that $-\nabla p = \nabla \cdot (u . \nabla u)$ and therefore $-\nabla p = F(u,u) + g$ where g is a harmonic bounded function ; therefore g is a constant vector. Now we deduce from the conservation of energy that $F(u,u)$ is bounded in L^1 by $C_6 \|u\|_{L^2} \cdot \|u\|_{0,\alpha}$. Since ∇p belongs to L^2 we have $g = 0$.

The uniqueness of the solution of (6) is easy to prove ; due to lemma 2 the only thing to prove is the existence of the solution of (6) and the fact that when $u_0 \in L^2$ u remains in L^2 . We start with $v \in C(R ; C^{1,\alpha})$ and $u_0 \in C^{1,\alpha}_\sigma$; with a truncation and a regularisation we can construct a sequence $u_0^R \in \mathcal{D}(R^3)$ uniformly bounded in $C^{1,\alpha}_\sigma$ going to u_0 in $C^{1,\alpha-\eta}$ strong $(0 < \eta < \alpha)$ when R goes to infinity; however u_0^R does not satisfies anymore the relation $\nabla . u_0^R = 0$. Now there exists a smooth function $\tilde{u}$ which is a solution of the equation

$$(27) \qquad \begin{cases} \frac{\partial \tilde{u}}{\partial t} + v.\nabla \tilde{u} = -\nabla p - v.\nabla u_0^R \ , \quad \nabla p \in L^2(R^3) \ , \\ \nabla . \tilde{u} = 0 \ , \quad \tilde{u}(.,0) = 0 \end{cases}$$

Next $u_R = \tilde{u} + u_0^R$ is the solution of

$$(28) \qquad \frac{\partial u_R}{\partial t} + v.\nabla u_R = -\nabla p \ , \quad u_R(.,0) = u_0 \quad .$$

Since $\frac{\partial}{\partial t}(\nabla . u_R) = 0$ we obtain :

$$(29) \qquad \nabla \cdot (v.\nabla u_R) = -\nabla p \quad \text{or} \quad -\nabla p = F(u_R, v) \quad .$$

Next using lemma 1 and lemma 2 we have

$$(30)\quad \frac{d}{dt_+}\|u_R\|_{0,\alpha} \le \|\nabla v\|_0 \|u_R\|_{0,\alpha} + C_7\|v\|_{0,\alpha}\|u_R\|_{1,\alpha}$$

and taking the curl of both sides of (28) we have

$$(31)\quad \frac{d}{dt_+}\|\nabla\wedge u_R\|_{0,\alpha} \le \|\nabla v\|_0\|\nabla\wedge u_R\|_{0,\alpha} + C_8\|v\|_{1,\alpha}\|u_R\|_{1,\alpha} \quad .$$

Adding (30) and (31) and noticing that $\frac{d}{dt_+}\|\nabla . u_R\|_{0,\alpha} = 0$, we obtain (here the term $\|u_R\|_{0,\alpha}$ appears because Ω is unbounded) :

$$(32)\quad \frac{d}{dt_+}\{\|\nabla\wedge u_R\|_{0,\alpha} + \|\nabla . u_R\|_{0,\alpha} + \|u_R\|_{0,\alpha}\}$$

$$\le \|\nabla v\|_0(\|u_R\|_{0,\alpha} + \|\nabla\wedge u_R\|_{0,\alpha}) + C_9\|v\|_{1,\alpha}\|u_R\|_{1,\alpha} \quad .$$

By the ellipticity there exists a constant C_{10} such that for any $w \in C^{1,\alpha}$

$$(33)\quad \|w\|_{1,\alpha} \le C_{10}\{\|w\|_{0,\alpha} + \|\nabla\wedge w\|_{0,\alpha} + \|\nabla . w\|_{0,\alpha}\} = C_{10}|||w|||_{0,\alpha} \quad .$$

Therefore we deduce from (32) and (33) the inequality :

$$(34)\quad \frac{d}{dt_+}|||u_R|||_{0,\alpha} \le \|\nabla v\|_0|||u_R|||_{0,\alpha} + C_{10}\|v\|_{1,\alpha}|||u_R|||_{0,\alpha} \quad .$$

This proves that u_R is bounded (uniformly in every compact of R) in $C^{1,\alpha}$. Now we let R go to infinity and it is easy to see that u_R will converge to the solution u of the equation .

$$(35)\quad \frac{\partial u}{\partial t} + v.\nabla u = F(u,v) \ , \quad \nabla . u = 0 \ , \quad u(.,0) = u_0(.) \ .$$

We therefore have proved that for every $v \in C(R \,;\, C^{1,\alpha})$ and $u_0 \in C^{1,\alpha}_\sigma$, equation (35) has a solution which satisfies the estimate

$$(36)\quad \frac{d}{dt_+}|||u|||_{0,\alpha} \le \|\nabla v\|_0|||u|||_{0,\alpha} + C_{10}\|v\|_{1,\alpha}|||u|||_{0,\alpha} \quad .$$

Now we can construct an iteration scheme ; we start with $u^0 = u_0$ we define u^n by the relation :

$$(37)\quad \frac{\partial u^n}{\partial t} + u^{n-1}.\nabla u^n = F(u^n \,,\, u^{n-1}) \ , \ \nabla . u^n = 0 \ , \ u^n(.,0) = u_0(.) \ ;$$

By (36) we have

$$(38) \qquad \frac{d}{dt_+} |||u^n|||_{0,\alpha} \leq C_2 |||u^{n-1}|||_{0,\alpha} \, |||u^n|||_{0,\alpha} \quad ,$$

and now for the rest of the proof we proceed as in the proof of theorem 1 .

The proof of proposition 1 is done by induction on k ; , for the sake of simplicity we give some details in the case of a bounded domain. Let u^n and ω^n be defined by (17) , assume that on the interval $|t| < T^* - \varepsilon$ $(\varepsilon > 0)$ $\|u^n\|_{k,\alpha}$ is uniformly bounded, then, if $u_0 \in C_\sigma^{k+1,\alpha}$, $\|u^n\|_{k+1,\alpha}$ is uniformly bounded on the same interval. Indeed, let $\ell = (\ell_1 , \ell_2 , \ell_3) \in N^3$ be a multiinteger of lenght $\ell_1 + \ell_2 + \ell_3 = k$, denote by D^ℓ the operator $\frac{\partial}{\partial x_1}^{\ell_1} \frac{\partial}{\partial x_2}^{\ell_2} \frac{\partial}{\partial x_3}^{\ell_3}$; with Leibnitz's rule we deduce from (17) the relation :

$$(39) \quad \frac{\partial}{\partial t} D^\ell \omega^{n+1} + u^n . \nabla (D^\ell \omega^{n+1}) = - \sum_{|p|=1} \binom{\ell-p}{p} D^p u^n . \nabla (D^{\ell-p} \omega^{n+1})$$

$$- \sum_{1<|p|,p\leq \ell} \binom{\ell-p}{p} D^p u^n . \nabla D^{\ell-p} \omega^{n+1} + D^\ell \omega^{n+1} . \nabla u^n$$

$$+ \sum_{0\leq p < \ell} \binom{\ell-p}{p} D^p \omega^{n+1} \nabla^{\ell-p} u^n \quad .$$

Now if u^n is uniformly bounded by K in $C^{k,\alpha}$ for $|t| < T - \varepsilon$ we can estimate the right side of (39) in $C^{0,\alpha}$. The first and third terms arc bounded by $C_{11} \|u^{n+1}\|_{k+1} \|u^n\|_{1,\alpha}$, while the second an the fourth are bounded by $C_{12} \|u^n\|_{k,\alpha} \|u^{n+1}\|_{k,\alpha} \leq C_{12} K^2$; (some analogous computation in Sobolev spaces can be found in [6]) therefore using the lemma 1 we have

$$(40) \quad \frac{d}{dt_+} \|D^\ell \omega^{n+1}\|_{0,\alpha} \leq C_{11} \|u^{n+1}\|_{k+1} \, \|u^n\|_{1,\alpha} + C_{12} K^2 \quad ,$$

for every ℓ $(|\ell| = k)$ now we add these inequations for $|\ell| = k$, and we obtain, using (4)

$$(41) \quad \frac{d}{dt_+} |||u^{n+1}|||_{k,\alpha} \leq C_{13} \, |||u^{n+1}|||_{k,\alpha} + C_{14}$$

This, with Gronwall's lemma, proves that u^n is uniformly bounded (for $|t| < T - \varepsilon$) in $C^{k+1,\alpha}$.

Finally we give the proof of proposition 2 .

Let u be the solution of the elliptic system $\nabla . u = 0$, $\nabla \wedge u = \omega$, $u . \nu|_{\partial\Omega} = 0$, then it is known that when $\omega \in L^{\infty}(\Omega)$ u generally does not belong to $W^{1,\infty}(\Omega)$ and is not even Lipschitzian. However, one has, when ω is bounded, the a priori estimate :

$$(42) \quad |u(x) - u(y)| \leq D(\Omega)\,|x-y|\,\mathrm{Log}\,\frac{L}{|x-y|}\,\|\nabla \wedge u\|_0$$

Therefore using the Lagrangian coordinates, one obtains for the solution of the Euler equation the a priori estimate

$$(43) \quad \rho_u'(x,y,t) = |u(x(t),t) - u(y(t),t| \leq D(\Omega)\,\rho_u(x,y,t)\log(L/\rho_u(x,y,t))\,\|\nabla\wedge u(t,.)\|_0$$

To obtain (8) we just integrate the inequality

$$(44) \quad \frac{\rho'}{\rho\,\mathrm{Log}(\frac{L}{\rho})} \leq D(\Omega)\,\|\nabla \wedge u(t,.)\|_0 \quad .$$

REFERENCES

[1] C. BARDOS, U.FRISCH, P.PENEL & P.L.SULEM , Modified dissipativity for a non linear evolution equation arising in Turbulence (1975)(these proceedings)

[2] S. BAOUENDI & C. GOULAOUIC (1975) (Private communication)

[3] S.ORSZAG, Lectures on the statistical theory of Turbulence; les Houches (1973)

[4] J.P. BOURGUIGNON & H. BREZIS , Remarks on the Euler Equation. J. Funct. Analysis - 15 - (1974) , pp.341-363 .

[5] D. EBIN & J.MARSDEN , Groups of Diffeomorphisms and the motion of an incompressible Fluid , Ann. of Math. 92, (1970), pp. 102-163 .

[6] U.FRISCH, C.FOIAS & R. TEMAM , Existence des solutions C^{∞} des équations d'Euler - C.R.A.S. 280 (24 Février 1975) , série A.505-508.

[7] T. KATO , On the classical solution of the two dimensional, non stationary Euler Equation , Arch. for . Rat. Mech. and Analysis 25 (3) (1967) pp. 188-200 .

[8] T. KATO , Non stationary flows of viscous and ideals fluids in $\mathbb{R}^3$ - J. Funct. Anal. (1972) pp.296-305.

[9] O.A LADYZENSKAIA & N. URALTCEVA , The mathematical theory of viscous incompressible flow. Gordon and Breach - New-York (1969) .

[10] M.LESIEUR & P.L.SULEM . Les équations spectrales en turbulence homogène et

isotrope;quelques résultats théoriques et numériques (1975) (these proceedings).

[11] L.LICHTENSTEIN , Uber einige Existenzprobleme der Hydrodynamik ... Math Z, 23 (1925) .

[12] R. TEMAM , On the Euler Equation of incompressible perfect fluids - Jour. Funct. Analysis, 20 (1975) , pp. 32-43.

[13] M. ZERNER, Sur une inegalité de Poincaré, (1975) to appear.

MODIFIED DISSIPATIVITY FOR A NON LINEAR EVOLUTION EQUATION ARISING IN TURBULENCE

C. Bardos
Dept. de Mathématique
IMSP, Université de Nice

P. Penel
Dept. de Mathématique
Université de Paris-Sud, Orsay

U. Frisch and P.L. Sulem
Centre National de la Recherche Scientifique
Observatoire de Nice

ABSTRACT

We are concerned with the global (in time) regularity properties of the Burgers MRCM equation, which arises in the theory of turbulence (with $\alpha = 1$)

$$\frac{\partial U}{\partial t}(t,x) = -\frac{\partial^2}{\partial x^2}[U(t,0) - U(t,x)]^2 - \nu\left(-\frac{\partial^2}{\partial x^2}\right)^{\alpha} U(t,x)$$

where $U(t,\cdot)$ is of positive type and where the dissipativity α is a nonnegative real number. It is shown that for arbitrary $\nu > 0$ and $\varepsilon > 0$, there exists a global solution in $L^{\infty}[0,\infty; H^{\frac{3}{2}-\varepsilon}(\mathbb{R})]$. If $\nu > 0$ and $\alpha > \alpha_{cr} = \frac{1}{2}$, smoothness of initial data persists indefinitely. If $0 < \alpha < \alpha_{cr}$, there exist positive data-dependent constants $\nu_1(\alpha)$ et $\nu_2(\alpha)$ such that indefinite persistence of regularity holds for $\nu > \nu_1(\alpha)$, whereas for $0 < \nu < \nu_2(\alpha)$ the second spatial derivative at the origin blows up after a finite time. It is conjectured that with a suitable choice of α_{cr}, similar results hold for the Navier-Stokes equation.

1. INTRODUCTION

We consider an equation which arises in the theory of homogeneous turbulence. "Homogeneous" as used by physicists means that the velocity field is a random function, the statistical properties of which are invariant under space translations. A contracted description of the random velocity field is given by its moments (if they exist), but one has to face a closure problem : it is not possible to derive from the Navier Stokes equation a deterministic system of n (finite) equations for n moments (Orszag 1975). Nevertheless, closed non linear equations for the velocity covariance are commonly found in the litterature on turbulence. They may be obtained on phenomonologic arguments by introducing a supplementary but arbitrary closure assumption. Kraichnan (1961) has shown that the resulting equations are also sometimes exact consequences of certain stochastic models (Herring and Kraichnan 1972).

As a simple example of the equations arising from closure, we shall consider the Burgers-MCRM equation

$$\frac{\partial U}{\partial t}(t,x) - \nu \frac{\partial^2 U}{\partial x^2}(t,x) + \frac{\partial^2}{\partial x^2}[U(t,0) - U(t,x)]^2 = 0$$

obtained by applying the Markovian Random Compling Model (Frisch et al. 1974) to the ordinary equation of Burgers (1940). In the above equation, $U(t,x)$ is defined as the mean product of velocities at the same time t and at two points separated by x , cleary a function of positive type in x .

The Burgers MRCM equation has been studied first heuristically by Brissaud et al. (1973) and then, from a mathematical view-point by Brauner, Penel and Temam (1974), Penel (1975), Foias and Penel (1975). The main results obtained previously are : in the inviscid case ($\nu = 0$), the solution remains as smooth as the initial data and energy is conserved up to a time t_* (depending on initial data) after which the second derivative at the origine blows up and energy is dissipated (energetic catastrophe) ; for $\nu > 0$, smoothness persists forever.

Our main interest in this paper will be in the problem of modified dissipativity when $\nu \frac{\partial^2}{\partial x^2}$ is replaced by $-\nu(-\frac{\partial^2}{\partial x^2})^\alpha$ where the power of the negative Laplacian, α , called dissipativity , is a non negative real number. Modified dissipativity in connection with the Navier-Stokes equation is considered in Ladyzenskaya (1963) and Lions (1969) where it is shown that global regularity holds for $\alpha > 5/4$. Frisch (1974) has conjectured that global regularity for the Navier Stokes equation with arbitrary $\nu > 0$ holds only for $\alpha > \alpha_{cr}$ with $0 < \alpha_{cr} < 1$. It will be shown in this paper that the modified dissipativity problem can be completely solved for the Burgers MRCM equation. It is easily seen that for this equation the critical value α lies indeed between zero and one : for $\alpha = 1$, we already know that global regularity holds ; for $\alpha = 0$, the change of variable $U(x,t) = e^{-\nu t} V(x,t)$ and $t' = \frac{1-e^{-\nu t}}{\nu}$ reduces the problem to the inviscid case for which we know that the second derivative at the origin blows up at $t' = t_*$, corresponding to a time $t = -\frac{1}{\nu} Ln(1-\nu t_*)$ which will actually be attained iff $\nu < \frac{1}{t_*}$.

The technical aspects of our study can be outlined as follows. The existence theorem is established by the Galerkin's method using an L^2 a priori estimate. A global regularity result uniform in the dissipativity and the viscosity is proved from an L^1 a priori estimate for the second derivative using a method of Kruzkov (1970) ; this "optimal" regularity is not strong enough to ensure uniqueness. The main theorem which gives the critical dissipativity $\alpha_{cr} = \frac{1}{2}$ has two aspects. First, to show global regularity for $\alpha > \frac{1}{2}$, we use a priori estimates in the form of differential inequations which are the analog of Kato's (1972) inequations for the Sobolev norms of the solutions of the Euler or Navier-Stokes equation. Second, to show loss of regularity for $0 < \alpha < \frac{1}{2}$ and sufficiently small viscosity, we use a lemma given by Foias and Penel (1975) and an estimate for the Fourier transform of the solution.

Only the main a priori estimates and the statement of the principal theorems are given here. A detailed version is forthcoming.

In the last chapter, we discuss the physical meaning of our results and show how some of the techniques can be carried over (in part) to the Navier-Stokes MRCM equation, and finally, to the true Navier-Stokes equation.

2. STATEMENT OF THE PROBLEM AND NOTATIONS

We are dealing with the Burgers MRCM equation

$$\frac{\partial U}{\partial t}(t,x) + \nu\left(-\frac{\partial^2}{\partial x^2}\right)^{\alpha} U(t,x) + \frac{\partial^2}{\partial x^2}[\,U(t,0) - U(t,x)\,]^2 = 0 \tag{1}$$

The function U is defined for $x \in \mathbb{R}$ and $t \in \mathbb{R}_+$ and satisfies the boundary and initial conditions

$$\lim_{|x|\to\infty} U(t,x) = 0 \tag{2}$$

$$U(0,x) = U_0(x) \tag{3}$$

The viscosity ν and the dissipativity α are non negative parameters. Because of the probabilistic origin of the problem, we shall be interested only in real solutions of positive type in x , i.e. which can be written

$$U(\cdot,x) = \int_{-\infty}^{+\infty} e^{-ikx}\ \hat{U}(\cdot,k)\ dk \tag{4}$$

where the Fourier transform in space, $\hat{U}(\cdot,k)$, called the <u>energy spectrum</u>, is positive and even.

Many a priori estimates of this paper make use of the <u>energy moments</u> defined, for $s \geq 0$, by

$$|U(t)|_s = \int_{-\infty}^{+\infty} k^{2s}\,\hat{U}(t,k)\ dk \tag{5}$$

$|U(t)|_0$ is called <u>energy</u> and $|U(t)|_1 = -\frac{\partial^2}{\partial x^2} U(t,0)$ is called <u>enstrophy</u> .
$\mathcal{E}(\mathbb{R})$ denotes the intersection of all the Sobolev spaces for $s \geq 0$, and B.V.$(\mathbb{R})$, the space of functions of bounded variations.

3. EXISTENCE AND GLOBAL REGULARITY THEOREM

The construction of a solution of the Burgers MRCM equation is done via an elliptic-regularization which reads (for $\mu > 0$)

$$\frac{\partial\bar{U}}{\partial t}(t,x) + \nu\left(-\frac{\partial^2}{\partial x^2}\right)^{\alpha}\bar{U}(t,x) - \mu\frac{\partial^2\bar{U}}{\partial x^2}(t,x) - 2\frac{\partial}{\partial x}\left[(\bar{U}(t,0) - \bar{U}(t,x))\frac{\partial\bar{U}}{\partial x}(t,x)\right] = 0 \tag{6}$$

$$\lim_{|x|\to\infty} \bar{U}(t,x) = 0 \tag{7}$$

$$\bar{U}(x,0) = U_0(x) \tag{8}$$

<u>An L^2 a priori estimate</u> : Let U be a sufficiently smooth solution of (1)-(3), then for arbitrary $\nu \geq 0$, $\alpha \geq 0$ and every $t \geq 0$,

$$\frac{\partial}{\partial t}\int_{-\infty}^{+\infty}[U(t,x)]^2dx + 2\nu\int_{-\infty}^{+\infty} U(t,x)\left(\frac{-\partial^2}{\partial x^2}\right)^{\alpha}U(t,x)dx + 4\int_{-\infty}^{+\infty}[U(t,0)-U(t,x)]^2\left[\frac{\partial U}{\partial x}(t,x)\right]^2dx=0 \tag{9}$$

<u>Existence for the regularized problem</u> : Let initial data U_0 of positive type be given in $\mathcal{E}(\mathbb{R})$. For arbitrary $\mu > 0$, there exists a unique positive type function $\bar{U}(t,x)$ solution of eqs. (6)-(8) in $\mathcal{C}^{\infty}[\mathbb{R}+, \mathcal{E}(\mathbb{R})]$.

<u>L^1 a priori estimates</u> : Let $\bar{U}$ be a sufficiently smooth solution of (6)-(8), then for any $t \geq 0$

$$\|\bar{U}(t)\|_{L^1(\mathbb{R})} \leq \|U_0\|_{L^1(\mathbb{R})} \tag{10}$$

$$\hat{\bar{U}}(t,k) \leq \frac{1}{2\pi}\|U_0\|_{L^1(\mathbb{R})} \tag{10'}$$

$$\left\|\frac{\partial^2\bar{U}}{\partial x^2}(t)\right\|_{L^1(\mathbb{R})} \leq \left\|\frac{d^2U_0}{dx^2}\right\|_{L^1(\mathbb{R})} \tag{11}$$

$$\hat{U}(t,k) < \frac{k^{-2}}{2\pi} \left\| \frac{d^2U_o}{dx^2} \right\|_{L^1(\mathbb{R})} \qquad (11')$$

Existence and optimal global regularity theorem for the non-regularized problem

Given $U_o \in \mathcal{E}(\mathbb{R})$, there exists for any $\nu \geq 0$ and $\alpha \geq 0$, and for any $\varepsilon > 0$, a weak solution U of (1)-(3) with

$$U \in L^\infty[\mathbb{R}+, \mathcal{C}^\infty(\mathbb{R})] \cap L^\infty[\mathbb{R}+, H^{\frac{3}{2}-\varepsilon}(\mathbb{R})] \qquad (12)$$

$$\frac{\partial U}{\partial x} \in L^\infty[\mathbb{R}+, L^\infty(\mathbb{R})] \cap L^\infty(\mathbb{R}+, \text{B.V.}) \qquad (13)$$

$$\hat{U}(t,k) \leq \frac{1}{2\pi} \|U_o\|_{L^1(\mathbb{R})} \qquad (14)$$

$$\hat{U}(t,k) \leq \frac{k^{-2}}{2\pi} \left\| \frac{\partial^2 U_o}{\partial x^2} \right\|_{L^1(\mathbb{R})} \qquad (15)$$

4. CRITICAL DISSIPATIVITY

Lemma (Foias and Penel 1975) : Let v a real function of positive type belonging to $H^1(\mathbb{R})$, then

$$\int_{-\infty}^{+\infty} [v(0)-v(x)][\frac{dv}{dx}(x)]^2\, dx \geq \frac{\pi}{2} q[\int_q^\infty \hat{v}(p)dp]^3 \qquad \forall q \geq 0 \qquad (16)$$

Theorem : Let U a solution given by the optimal global regularity theorem, then

$$|U(t)|_s \leq \sqrt{2}\, |U_o|_o + \frac{1}{\pi(1-2s)} \left\| \frac{d^2U_o}{dx^2} \right\|_{L^1(\mathbb{R})}, \qquad \text{for } 0 \leq s < 1/2 \qquad (17)$$

$$\frac{d}{dt} \|U(t)\|^2_{L^2(\mathbb{R})} \leq -\pi q \{ |U(t)|_o - \frac{q}{\pi} \|U_o\|_{L^1(\mathbb{R})} \}^3, \qquad \forall q \geq 0 \qquad (18)$$

Hölder type lemma : Let $0 \leq s_1 \leq s_2 \leq s_3$, and f be a function of positive type such that $|f|_{s_1} < +\infty$ and $|f|_{s_3} < +\infty$, then

$$|f|_{s_2}^{s_3-s_1} \leq |f|_{s_1}^{s_2-s_3}\, |f|_{s_3}^{s_2-s_1} \qquad (19)$$

A priori estimates for the energy moments : Let U be a positive type solution of (1)-(3) belonging to $\mathcal{C}^\infty([0,T[, \mathcal{E}(\mathbb{R}))$, then for arbitrary $\nu \geq 0$, and $\alpha \geq 0$, and for any $t \in [0,T[$,

(i) the energy equation reads *

$$\frac{d}{dt} |U(t)|_0 + \nu |U(t)|_\alpha = 0 \qquad (20)$$

(ii) the enstrophy equation reads

$$\frac{d}{dt} |U(t)|_1 + \nu |U(t)|_{1+\alpha} = 6\, |U(t)|_1^2 \qquad (21)$$

(iii) There exists a positive constant λ_s such that, for $s > 0$

$$\frac{d}{dt} |U(t)|_s + \nu |U(t)|_{\alpha+s} \leqslant \lambda_s |U(t)|_1 |U(t)|_s \qquad (22)$$

(iv) For integer $n > 1$,

$$\frac{d}{dt} |U(t)|_n + \nu |U(t)|_{n+\alpha} \geqslant (2n+1)(2n+2) |U(t)|_1 |U(t)|_n \qquad (23)$$

Loss of analyticity : From (23), it can be shown that for $\nu = 0$ and analytic initial data U_0, analyticity may be lost for any $t > 0$. This is at variance to the ordinary Burgers and Euler equations (Goulaouic and Baouendi 1975).

Main theorem : Let initial data of positive type be given in $\mathscr{E}(\mathbb{R})$ and consider the solution $t \to U(t)$ given by the existence and global regularity theorem, then

i) for any $\nu > 0$ and $\alpha > 1/2$, the solution belongs to $\mathscr{C}^\infty[\mathbb{R}+, \mathscr{E}(\mathbb{R})]$ and there is uniqueness in this class of functions

ii) for $0 \leqslant \alpha < \frac{1}{2}$, there exist positive ν_1 and ν_2 $(\nu_1 \geqslant \nu_2)$ depending on α and on the initial data such that

(ii') for $\nu \geqslant \nu_1$, the same conclusion as in (i) holds ;

(ii") for $0 \leqslant \nu < \nu_2$, there exist t_* (depending only on initial data) and t_{**} ($> t_*$ and depending on α, ν and initial data) such that

- for $0 \leqslant t < t_*$, the same conclusion as in (i) holds,
- for $t > t_{**}$, $\mathscr{C}^1$ regularity in space breaks down and the energy equation holds in the form (20') instead of (20). As a consequence, $k^2 U(t,k)$ has a non zero limit for $k \to \infty$. The uniqueness problem remains open.

----

* The energy equation (20) requires $\mathscr{C}^1$ regularity in space. When this does not hold, energy is removed at a rate superior to what can be accounted for solely by the dissipative term (energy catastrophe) ; the energy equation then reads (Penel 1975)

$$\frac{d}{dt} |U(t)|_0 + \nu |U(t)|_\alpha + 2[\lim_{x \to 0^\pm} \frac{\partial U}{\partial x}(t,x)]^2 = 0 \qquad (20')$$

Remark : It is an open problem whether for $0 < \alpha < \frac{1}{2}$, one can take $\nu_1 = \nu_2$ and $t_* = t_{**}$. Another open problem is the behaviour of the solution for the critical dissipativity $\alpha_{cr} = \frac{1}{2}$.

5. PHYSICAL INTERPRETATION AND MATHEMATICAL PERSPECTIVES.

5.1 The physics of energy transfer

The energy spectrum $\hat{U}(t,k)$ describes the way energy is distributed among the various scales of motion. Kolmogorov's (1941) phenomenological theory predicts that for the Navier-Stokes equation when $\nu \to 0$, the spectrum should go like $k^{-5/3}$ for $k \to \infty$. The corresponding argument for the Burgers MRCM equation predicts k^{-2} (Frisch et al. 1974) . From chapter 2 , we know that $\hat{U}(t,k) \lesssim Ck^{-2}$; if the initial spectrum decreases faster than k^{-2} , the k^{-2} regime (in the limit $\nu \to 0$) will be established only after a finite time (Brissaud et al. 1973, Penel 1975).

To understand the meaning of the results on modified dissipativity, it is useful to think of two competing effects : first, the non-linear term does not change the total energy but tends to transfer it to ever larger wavenumbers therefore, increasing, possibly to infinity, the space derivatives ; second, the dissipative term $[-\nu k^{2\alpha} \hat{U}(t,k)]$ in Fourier representation extracts energy at a rate which increases with increasing wavenumber (if $\alpha > 0$) . For global regularity to hold, the dissipative term must be able to cope with non linear transfer. Why this requires $\alpha > \frac{1}{2}$ may be heuristically understood as follows. The rate of energy dissipation is $\nu \int_{-\infty}^{+\infty} k^{2\alpha} \hat{U}(t,k)dk$; if $\alpha < \frac{1}{2}$ and because $\hat{U}(t,k) \lesssim Ck^{-2}$, the integral will be uniformly bounded in ν and dissipation will go to zero with ν ; it is therefore not possible any more to remove energy which rushes to infinite wavenumbers. For true turbulence, according to present ideas, the energy spectrum should go like k^{-n} where n is slightly in excess of 5/3 (Kolmogorov 1962, Kraichnan 1974). The above heuristic argument gives then a critical dissipativity for the Navier-Stokes equation $\alpha_{cr} = \frac{n-1}{2}$ which should be close to $\frac{1}{3}$.

5.2 The MRCM Navier-Stokes equation

The procedure which leads from the Burgers equation to the Burgers MRCM equation may be applied to the Navier Stokes equation giving the Navier Stokes MRCM equation for the energy spectrum of homogeneous isotropic turbulence (Frisch et al.

1974, Lesieur and Sulem 1975).

$$\frac{\partial E}{\partial t}(t,k) + \nu k^{2\alpha} E(t,k) = \frac{1}{4} \iint_{\Delta_k} \frac{k}{pq}\, b_{kpq} \left[k^2 E(t,p)E(t,q) - p^2 E(t,q)E(t,k)\right] dpdq \qquad (24)$$

where Δ_k denotes the domain in the (p,q)-plane such that k,p,q can form a triangle and

$$b_{kpq} = \frac{p}{k}(xy + z^3) \quad ,$$

x, y, and z being the cosines of the interior angles of the (k,p,q)-triangle. Modified dissipativity has been assumed.

Contrary to the Burgers MRCM equation (with $\alpha = 1$), eq. (24) remains integrodifferential when it is written in position-space ; this follows form the presence of a non local pressure term in Navier Stokes equation.

Several a priori estimates established for Burgers MRCM are still valid for (24). Defining

$$|E(t)|_s = \int_0^\infty k^{2s} E(t,k)dk \quad ,$$

we have for smooth enough solutions, the energy equation

$$\frac{d}{dt}|E(t)|_0 + \nu|E(t)|_\alpha = 0 \qquad (25)$$

and the enstrophy equation (Lesieur 1973)

$$\frac{d}{dt}|E(t)|_1 = \nu|E(t)|_{1+\alpha} = \frac{2}{3}\; |E(t)|_1^2 \qquad (26)$$

From these a priori estimates, one can deduce a global existence, unique ness and smoothness theorem for Navier Stokes MRCM with the usual dissipativity ($\alpha = 1$) and arbitrary positive viscosity. The essential step of the proof is given below. Using the Schwarz inequality

$$|E(t)|_2 \geqslant |E(t)|_1^2 \;/\; |E(t)|_0 \quad ,$$

we obtain

$$\frac{d}{dt}|E(t)|_1 \leqslant |E(t)|_1 \{ \frac{2}{3}|E(t)|_1 - \nu|E(t)|_1 \, |E(t)|_0^{-1} \} \quad .$$

Upon integration and use of (25), we obtain

$$|E(t)|_1 \leqslant |E(0)|_1 \,.\, \exp\,[\frac{2}{3\nu}\,|E(0)|_0\,] \qquad (27)$$

which establishes the global boundedness of the enstrophy.

The question of critical dissipativity for Navier-Stokes MRCM is not settled. Numerical results have been obtained which indicate that $\alpha_{cr} = \frac{1}{2}$. The crucial estimate which is missing is the analog of $\hat{U}(t,k) \leq Ck^{-2}$. For Burgers MRCM equation, this was derived from an L^1 estimate in position-space. The presence of the pressure term makes now the problem more difficult. An alternative would be to obtain a proof of $\hat{U}(t,k) \leq Ck^{-2}$ for Burgers MRCM by working entirely in Fourier space and then to carry over to Navier-Stokes MRCM.

5.3 Possible extension to the Navier-Stokes equation ?

It is of interest to try to extend some of the techniques of this paper to the Navier Stokes equation in $\mathbb{R}^3$.

$$\frac{\partial v}{\partial t} + (v.\nabla)v - \nu(-\Delta)^{\alpha} v = -\nabla p \qquad \nabla . v = 0 \tag{28}$$

First one checks that global regularity properties are the same if the dissipative term is replaced by $-\nu(I-\Delta)^{\alpha}$ where I is the identity. One then has the following a priori estimate

$$\frac{d}{dt} ||v||_s^2 + 2\nu ||v||_{s+\alpha}^2 \leq C_s ||V||_s^3 \qquad s > 5/2 \tag{29}$$

where $||v||_s$ denotes the Sobolev norm in the subspace of $[H^s(\mathbb{R}^3)]^3$ of incompressible functions. For $s \geq 3$, (29) is a trivial extension of Kato's (1972) results ; to show that it holds for $s > 5/2$ requires a more defined analysis which will be published elsewhere. (29) is the exact analog of (22) for Burgers MRCM. The Hölder type inequality (19) still holds if $|U|_s$ is changed into $||v||_s$; therefore the argument of chapter 4 can be carried over to the Navier Stokes equation. Using the a priori energy estimate $||v(t)||_0 \leq ||v(0)||_0$, one establishes global regularity for $\alpha > 5/4$, which is not any better than Lions' (1969) result, except that a bounded set is not assumed. Why is there such a gap between this result and the corresponding one for Burgers MRCM ? Clearly, because for Burgers MRCM, we have a much better a priori estimate than the energy estimate, namely $\left|\left|\frac{\partial^2 U}{\partial x^2}(t)\right|\right|_{L^1(\mathbb{R})} \leq \left|\left|\frac{\partial^2 U}{\partial x^2}(0)\right|\right|_{L^1(\mathbb{R})}$ which for $\alpha > \alpha_{cr}$ and $\nu > 0$ is uniform in ν and t and for $\nu = 0$ and $0 \leq t \leq t_*$ is uniform in t . This suggests that improvement of the 5/4 result and particularly a proof of global regularity for the Navier Stokes equation with the usual dissipativity require essentially better a priori estimates for the Euler equation.

REFERENCES

BRAUNER, C.M., PENEL, P. and TEMAM, R. (1974) C.R. Acad. Sc. Paris, A.279, 65 and 115.

BRISSAUD, A., FRISCH, U., LEORAT, J., LESIEUR, M., MAZURE, A., POUQUET, A., SADOURNY, R., and SULEM, P.L., (1973), Ann. Geophys., 29 , 539.

BURGERS, J.M. (1940), Proc. Roy. Netherl. Acad., 43 , 2.

FOIAS, C. and PENEL, P. (1975), C.R. Acad. Sc. Paris, A.280, 629.

FRISCH, U. (1974), Proceedings of the Conference on Prospect for Theoretical Turbulence Research NCAR, Boulder, Colorado.

FRISCH, U., LESIEUR, M., and BRISSAUD, A. (1974), J. Fluid Mech., 65, 145.

GOULAOUIC, C. and BAOUENDI, S. (1975), Private Communication.

HERRING, J.R. and KRAICHNAN, R.H. (1972) in Statistical Models and Turbulence, p. 148, Springer.

KATO, T. (1972), J. Funct. Anal., 9 , 296.

KOLMOGOROV, N.A. (1941), C.R. Acad. Sc. URSS, 30, 301.

KOLMOGOROV, N.A. (1962), J. Fluid Mech., 12 , 82.

KRAICHNAN, R.H. (1961), J. Math. Phys., 2 , 124 ; also, erratum 3 , 205 (1962).

KRAICHNAN, R.H. (1974), J. Fluid Mech., 62 , 305.

KRUZKOV, S.N. (1970), First Order Quasilinear Equations in Several Independent Variables. Math. USSR Sbornik, vol. 10, 217.

LADYZENSKAYA, O.A. (1963), A Mathematical Theory of Viscous Incompressible Flow. (First edition, Gordon and Breach, New-York).

LESIEUR, M. (1973), Thesis, University of Nice.

LESIEUR, M. and SULEM, P.L. (1975), Les Equations Spectrales en Turbulence Homogène et Isotrope. Quelques Résultats Théoriques et Numériques. Proc. of this Conference.

LIONS, J.L. (1969), Quelques Méthodes de Résolution des Problèmes aux Limites non Lineaires. Dunod-Gauthier-Villars.

ORSZAG, S.A. (1975), Lectures on the Statistical Theory of Turbulence. Proceedings of the 1973 Les Houches Summer School of Theoretical Physics.

PENEL, P. (1975), Thesis, University of Paris-Sud, Orsay.

A GENERIC PROPERTY OF THE SET OF STATIONARY SOLUTIONS OF NAVIER STOKES EQUATIONS

C. FOIAS and R. TEMAM

This is a preliminary report on a progressing work devoted to generic properties of Navier Stokes equations. This Note deals with the generic finiteness of the set of stationary solutions.

Let Ω be a bounded open set in $\mathbb{R}^n$, $n = 2$ or 3, with a boundary Γ of class $\mathscr{C}^2$, and let us introduce the usual spaces [6]-[8]

$$\mathscr{D}(\Omega) = \text{space of real } \mathscr{C}^\infty \text{ functions with compact support in } \Omega ,$$

$$\mathscr{V} = \{u \in \mathscr{D}(\Omega)^n, \ \operatorname{div} u = 0\} ,$$

$$V = \text{the closure of } \mathscr{V} \text{ in } H_0^1(\Omega)^n$$
$$= \{u \in H_0^1(\Omega)^n, \ \operatorname{div} u = 0\}$$

$$H = \text{the closure of } \mathscr{V} \text{ in } L^2(\Omega)^n$$
$$= \{u \in L^2(\Omega)^n, \ \operatorname{div} u = 0 , \ u.\nu = 0 \text{ on } \Gamma, \ \nu \text{ the unit outward normal}\} .$$

We denote by (u,v) the scalar product in $L^2(\Omega)^n$ and H, and by $((u,v))$ the scalar product in $H_0^1(\Omega)^n$ and V :

$$((u,v)) = \sum_{i,j=1}^{n} \int_\Omega \frac{\partial u_i}{\partial x_j} \frac{\partial v_i}{\partial x_j} dx .$$

If V' denotes the dual of V, then as usual

$$V \subset H \subset V'$$

where the injection are dense, continuous and also compact.

We consider the stationary Navier Stokes problem with homogeneous boundary conditions : given f, to find u and p satisfying :

$$(1) \qquad - \nu \Delta u + \sum_{i=1}^{n} u_i \frac{\partial u}{\partial x_i} + \operatorname{grad} p = f \quad \text{in } \Omega ,$$

$$(2) \qquad \operatorname{div} u = 0 \quad \text{in } \Omega ,$$

(3) $$u = 0 \quad \text{on} \quad \Gamma .$$

It is well known (cf. Leray [5]) that this is equivalent to the following variational problem :

(4) To find $u \in V$ such that

$$\nu((u,v)) + b(u,u,v) = (f,v) , \quad \forall v \in V ,$$

where f is given in H .

Let A (or B) be the linear (or bilinear) continuous operator from V (or $V \times V$) into V' defined by

$$(Au,v) = ((u,v)) , \qquad \forall u,v \in V$$

$$(B(u,v),w) = b(u,v,w) = \sum_{i,j=1}^{n} \int_\Omega u_i \frac{\partial v_j}{\partial x_i} w_j \, dx , \qquad u,v,w \in V ,$$

$$B(u) = B(u,u) .$$

The equation (4) is equivalent to

(5) $$\nu Au + B(u) = f .$$

We denote by $S(f,\nu)$ the set of solutions $u \in V$ of (5). The following properties of $S(f,\nu)$ are well-known :

a. $S(f,\nu)$ is not empty (i.e. existence of solutions for (5)).

b. $S(f,\nu)$ reduces to one point (i.e. uniqueness of solution for (5)) if

(6) $$\nu^2 > c_0 |f| ,$$

where c_0 is a constant depending only on Ω .

c. $S(f,\nu) \subset H^2(\Omega)^n$ (i.e. regularity of the solutions of (5)), cf. [2]-[8]-[9].

We also have the following (cf. [3]).

d. $S(f,\nu)$ is compact in $H^2(\Omega)^n$, V and H .

Another property of $S(f,\nu)$ is the following one .

THEOREM 1. *For every fixed* $\nu>0$ *, there exists a dense* G_δ *subset* Σ *of* H *such that* $S(f,\nu)$ *is finite,* $\forall f \in \Sigma$.

The principle of the proof of the Theorem is as follows :

We consider the space $D(A) = A^{-1}H$ equiped with the Hilbert norm $|Au|$, which is equivalent to the norm induceed by $H^2(\Omega)^n$ $(D(A) \subset H^2(\Omega)^n)$. We consider the mapping

$$u \longmapsto \mathcal{E}(u) = \nu Au + B(u)$$

from $D(A)$ into H . This is a Frechet differentiable mapping with differential $\mathcal{E}'(u)$:

$$\mathcal{E}'(u).v = \nu Av + B(u,v) + B(v,u) \; , \quad \forall u, v \in D(A) \; .$$

Now assume that $S(f,\nu)$ is not finite : then there exists a sequence of mutually distinct elements $u_j \in S(f,\nu)$ which is convergent in $D(A)$ to some limit $u_o \in S(f,\nu)$ (cf.point d). By passage to the limit in the relation

$$\nu A(u_j - u_o) + B(u_j,u_j) - B(u_o,u_o) = 0$$

we find a non zero vector v ,

$$v = \lim \frac{u_j - u_o}{|u_j - u_o|} \; ,$$

such that

$$\mathcal{E}'(u_o).v = \nu Av + B(u_o,v) + B(v,u_o) = 0 \; .$$

Whence $\mathcal{E}'(u_o)$ is not regular (i.e. is not surjective) and $\mathcal{E}(u_o) = f$ is a singular value or critical point of $\mathcal{E}$ (cf.[1] [7]).It follows now from an infinite dimensional version of Sard Theorem proved in [1] [7] and which applies to $\mathcal{E}$(1) that the regular points of $\mathcal{E}$ (i.e. the non-singular points) constitute a dense G_δ subset of H , and the result follows.

This infinite dimensional version of Sard Theorem was recently indicated to us by M. Berger. Before being aware of this result, the authors completed a similar proof of Theorem 1 based only on the classical Sard Theorem, i.e. its finite dimensional version.

Let us sketch the principle of the first proof which keeps perhaps some interest.

Let w_j, λ_j, $j = 1,\ldots$ denote the eigenfunctions and eigenvalues of the

(1) For the details see [3].

operator A^{-1} which is compact and self adjoint in H. Let V_m be the space spanned by $w_1,\dots,w_m$, and P_m the projector in V or H on V_m.

We are able to prove the following technical result : for every $\nu>0$, for every $k>0$, there exists $m_1 = m_1(k,\nu)$ such that for each $g \in P_m H$, $|g| \leqslant k$ and for each integer $m \geqslant m_1$, the solution u of

$$(7) \qquad \mathcal{E}(u) = g ,$$

has a projection $Q_m u$ on V_m, which solely depends on $P_m u$ and the mapping $P_m u \longrightarrow u$ is one to one and analytic. For such a g, the equation (7) is equivalent to an equation

$$(8) \qquad \phi_m(\zeta) = g , \qquad \zeta \in \mathcal{O}_m ,$$

where $\zeta = P_m u$, $\mathcal{O}_m$ is some open set of V_m and ϕ_m is analytic from $\mathcal{O}_m$ into V_m.

By the classical Sard Theorem applied to ϕ_m, we obtain a set $B_m(k) \subset P_m B(k)$,

$$B(k) = \{f \in H, \ |f| < k\} ,$$

which is dense in $P_m B(k)$, $m \geqslant m_1(k,\nu)$, and such that $\mathcal{E}^{-1}(f)$ is finite $\forall f \in B_m(k)$. Whence

$$\Sigma = \bigcup_{k=1}^{\infty} \ \bigcup_{m \geqslant m_1(k,\nu)} B_m(k)$$

is dense in H and $\mathcal{E}^{-1}(f)$ is finite for every f in this set.

The same methods apply to the stationary nonhomogeneous Navier Stokes equations, and the problem of periodic solutions; also to many similar equations, see [3].

References

[1] R. Abraham, J. Robin -
Transversal mappings and flows.
W.A. Benjamin, New York, Amsterdam, 1967.

[2] L. Cattabriga -
Su un problema al contorno relativo al sistema di equazioni di Stokes.
Rend. Mat. Sem. Univ. Padova, vol.31, 1961, p.308-340.

[3] C. Foias, R. Temam - To appear.

[4] O.A. Ladyzhenskaya -
The mathematical theory of viscous incompressible flow.
Gordon and Breach, New York, 1969.

[5] J. Leray -
Etude de diverses équations intégrales non linéaires et de quelques problèmes que pose l'hydrodynamique.
J. Math. Pures Appl., vol.12, 1933, p.1-82.

[6] J.L. Lions -
Quelques méthodes de résolution des problèmes aux limites non linéaires.
Dunod-Gauthier-Villars, Paris, 1969.

[7] S. Smale -
An infinite dimensional version of Sard's Theorem.
Amer. J. Math., vol.87, 1965, p.861-866.

[8] R. Temam -
Navier-Stokes equations.
North-Holland-Elsevier, Amsterdam-New York, 1976.

[9] I.I. Vorovich, V.I. Yudovich -
Stationary flows of incompressible viscous fluids.
Mat. Sbornik, vol.53, 1961, p.393-428.

Université de Bucarest
Facultë de Mathématique
Rue Akademiei 14
Bucarest, Roumanie

Mathématiques
Université de Paris-Sud
91405 - Orsay, France

TWO STRANGE ATTRACTORS WITH A SIMPLE STRUCTURE

M. HENON[+] and Y. POMEAU[*]

[+]Observatoire de Nice

[*]DPh-T CEN SACLAY
BP n°2 - 91190 Gif-sur-Yvette, France

ABSTRACT

Numerical computations have shown that, for a range of values of the parameters, the Lorenz system of three non linear ordinary differential equations of first order has a strange attractor whose structure may be understood quite easily.

We show that the same properties can be observed in a simple mapping of the plane defined by : $x_{i+1} = y_i + 1 - a x_i^2$, $y_{i+1} = b x_i$. Numerical experiments are carried out for $a = 1.4$, $b = 0.3$. Depending on the initial point (x_0, y_0), the sequence of points obtained by iteration of the mapping either diverges to infinity or tends to a strange attractor, which appears to be the product of a one-dimensional manifold by a Cantor set. This strange attractor has basically the same structure than a plane section of the attractor found for the Lorenz system.

I - INTRODUCTION

Lorenz[1] proposed and studied a remarkable system of three coupled first-order differential equations, representing a flow in three-dimensional space. The divergence of the flow has a constant negative value, so that any volume shrinks exponentially with time. Moreover, there exists a bounded region R into which every trajectory becomes eventually trapped. Therefore, all trajectories tend to a set of measure zero, called attractor. In some cases the attractor is simply a point (which is then a stable equilibrium point) or a closed curve (known as a limit cycle). But in other cases the attractor has a much more complex structure. This is known as a strange attractor. Inside the attractor, trajectories wander in an apparently erratic manner. Moreover, they are highly sensitive to initial conditions.

All the known examples show that for differential systems of order 3 a "strange attractor" is an object which is intermediate "between" a surface in the ordinary sense and a volume : it may be viewed as a surface with an infinite number of sheets. As suggested by Thom[2] these strange attractors are continuous in some dimensions and have the structure of a Cantor ensemble in the remaining dimension : consider a point $\vec{x}$ on the attractor, then a local system of curvilinear coordinates exists such as if $\vec{x} = (0,0,0)$ in this system, thus the point of coordinates (u_1,u_2,u_3) is on the attractor when u_1 and u_2 vary in a finite interval around 0 and when u_3 is an element of Cantor set (or "Cantor like" set). These strange attractors have been already found in studying simple non linear differential equations related to various problems : the case d = 3 has been already encountered in studies on the unsteady Benard-Rayleigh thermo-convection[1] and on the reversals of the geomagnetic field[3]. As explained by Ruelle, the existence of strange attractors for so simple deterministic systems clearly demonstrates the possibility of randomness in phenomena as turbulence without any connection with the existence of an "infinite number of degrees of freedom". We present here two cases of strange attractors that have been found by studying the Lorenz system (section II) and then (section III) by trying to reproduce the Poincaré transform for this system by a planar quadratic transform.

II - THE STRANGE ATTRACTOR FOR THE LORENZ SYSTEM

II.A - The Lorenz system :transition from a strange attractor to a limit cycle

The Lorenz system is obtained by truncating the Oberbeck Boussinesq fluid equations for a layer heated from below. One keeps only a few spatial harmonics of the velocity and temperature field at fixed wavenumber . The three remaining variables, x_1,x_2 and x_3 are the amplitudes of the first spatial harmonics of the velocity and temperature fluctuations and of the zeroth harmonic of the temperature fluctuation. The time evolution of these three quantities is given by :

$$\begin{aligned} \dot{x}_1 &= \sigma(x_2 - x_1) \\ \dot{x}_2 &= -x_1x_3 + rx_1 - x_2 \\ \dot{x}_3 &= x_1x_2 - \beta x_3 , \end{aligned} \qquad (1)$$

where σ, r and β are numbers. Lorenz has shown[1] that, for $\sigma = 10$, $\beta = 8/3$ and $r = 28$, the point $\vec{x}(t)$ moves, after some transients, on a "strange attractor", which has been studied since by Landford and Ruelle[4]. We have integrated the system (1) on an analog computer by letting σ and β constant ($\sigma = 10$, $\beta = 8/3$) and making r vary from values close to the one chosen by Lorenz ($r = 28$) to higher ones. During this investigation it has appeared that, in a range of values of r, the motion had a strange attractor that looked quite simple. In this range, we have made more detailed investigations on a digital computer which will be reported here.

We had already observed on the analog computer that the Lorenz system has a number of sharply defined transitions between limit cycles in the usual sense (i.e. closed curves in the 3d space) and strange attractors. One of these transitions occurs around $r \simeq 220$. When r is slightly larger than the critical value, the trajectory is attracted by a closed curve (at least at the accuracy of the analog calculation which is not very easy to assess), which has been found also in a double precision digital computation and which is reproduced on (Fig.1). When r is smaller than the critical value, the trajectory lies on a "strange attractor" which, at first sight, looks like a surface. On Fig.2, we have reproduced two projections of this surface on the planes (x_1,x_2) and (x_2,x_3) respectively. Let us comment briefly upon a few points that appear at once from these figures.

i) In the domain of values of parameters under consideration, they are 3 unstable fixed points : $(0,0,0)$, $(\pm\sqrt{\beta(r-1)}, \pm\sqrt{\beta(r-1)}, r-1)$. The last two fixed points are plotted on Fig.2.

ii) In this domain of values of r the original symmetry of the Lorenz equations : $(x_1,x_2,x_3) \rightarrow (-x_1,-x_2,x_3)$ is spontaneously broken as neither the limit curve nor the strange attractor have any simple symmetry.

iii) The limit curve and the strange attractor look very similar and one may wonder wether the strange attractor may be build up by small instabilities around the limit curve. We shall come back to this point later on.

iv) In order to get a better idea of the shape of the attractor, we have drawn the projections of the intersect of this attractor with a set of parallel planes (Fig.3).

II.B - The Poincaré transform

On this attractor the motion runs always in a quite well definite direction, so that it is meaningful to follow the change of the sections along this motion, in order to understand the difference between the limit curve and the strange attractor. For that purpose we have found an approximately periodic trajectory on the attractor, we shall explain below how to get it from the consideration of the so called Poincaré transform.[5] Presumably this almost periodic trajectory denotes the presence of a periodic solution of the equations of the motion, it is very close to the stable limit cycles obtained for slightly different values of the parameters, but contrary to this limit cycle it is "a little" unstable. At series of points we have cut the surface by planes perpendicular to this periodic trajectory. These various sections are located by letters A,B,C,... on Fig.4 and they are drawn in the same order on Fig.5.

This Fig.5 makes the picture of the strange attractor quite clear. At the beginning an (apparently) simple section, as the A-section, as the A-section, is almost without any curvature ; when the mean point follows its trajectory, this section becomes curved as, say, a capital U with two unequal branches, then the two sides of this U go closer and closer to each other and, finally, the two sheets collapse, at least at the accuracy of the computation and one recovers the initial section. Of course, as already noticed by Lorenz, two sheets do never really merge, owing to the deterministic character of the equations of the motion, they just become nearer and nearer.

The behaviour of the successive sections as the motion goes on may be understood by drawing in the planes normal to the (unstable) periodic trajectory both the attractor (that appears as a line) and the vector field made by the projection on to this plane of the velocity field defined by (1). This is done in Fig.6. This attractor is made as follows : the fluctuations near the periodic trajectory are unstable in one direction, i.e. the 2d vector field in the normal plane is hyperbolic in the vicinity of the fixed point. Thus the surface is stretched along this direction. This occurs from section (K) to (A). It turns out that, between (approximately) sections (C) and (E) the unstable and stable direction in this normal plane rotate quite rapidly. The section of the attractor is no more directed along one of the principal axis of the vector field it is contracted and folds up partly when moving toward the attracting direction. At the end of the process the section has folded up and is again directed (approximately) along the unstable axis. Following continuously the end of one of the main directions of the motion in the normal plane near to the origin, one finds that, after one turn, this has turned (as does approximately this section itself) of an angle of Γ .

Now it is possible to deduce (qualitatively of course !) the structure of the attractor from what happens during a single run. Starting from an ensemble of initial conditions in a plane around the section A, one obtain after a complete run around the attractor a very thin U-shaped ensemble of arrival points. The correspondence between the starting and arrival points is, by definition, the Poincaré mapping. The cut of the attractor is the result of an infinite number of applications of this mapping to a suitably chosen ensemble of initial conditions.

As already noticed by Lorenz, the system (1) is contracting, i.e. $\sum_{i=1}^{3} \frac{\partial(x_i)}{\partial x_i} = -(\sigma+\beta+1)$ is negative, so that the volume of an ensemble of initial conditions is contracted by a factor $\rho = e^{-(\beta+\sigma+1)T} \simeq 1.07\times10^{-3}$ by one application of the Poincaré mapping, $T \simeq 0.5$ being the mean period of rotation. In order to make the reasoning simpler, let us assume for the moment that the two sheets generated form one sheet after one turn are approximately equal, and that their distance is δ (which is too small to be seen on our Fig.5). Starting from an ensemble initial conditions of thickness δ , one obtains after one turn two sheets of thickness $\delta\rho/2 (\rho \ll 1)$ that are distant of δ . After another turn, another sheet of relative thickness $(1-\rho)$ will be again deleted from the central part of these two new sheets, generating a 4 sheets ensemble, and so on (see Fig.7). Let us consider now the points located on a straight line approximately normal to the attractor. The initial conditions fill up a segment of length δ on this line ; after one turn the central part of this segment has been deleted, at the next turn the central part of each of these two small segments is again deleted, and so on. Of course this is the method for generating a Cantor ensemble.

This allows one to conclude that, in the present case, the section of the strange attractor by a straight line is (if it is not empty, of course), made of Cantor sets. This argument can be made more rigorous by constructing a simplified model of the Poincaré mapping. Our picture of this mapping is very close to the well known "baker's transform" except for an important difference : owing to the contraction by the factor ρ , the mapping induced by the Lorenz equation does not preserve the area, as the baker's transform does. Consider the baker's transform $B(x,y) \to (Bx,By)$, where (x,y) belongs to the unit square $[0,1]^2$, and where

$$Bx = 2x \ , By = y/2 \text{ if } 0 \le x < 1/2 \text{ and } Bx = 2x-1 \ By = (y+1)/2 \text{ if } 1/2 \le x \le 1 .$$

In order to get a model of the Poincaré transform generated in the present case by the Lorenz system, it suffices to compose this baker's transform with another transform, say C, that "concentrates" the density of arrival points on two horizontal sheets around, say $y = 1/4$ and $3/4$. This transform C may be taken as

$$C\ (x,y) \to (x,Cy) \ ,$$

where $Cy = 1/8 + y/2$ if $0 \le y < 1/2$ and $Cy = 3/8 + y/2$ if $1/2 < y \le 1$.

Of course, many other forms could be imagined for C. The most important property is that $y \to Cy$ has two stable fixed points : $y = 1/4$ and $3/4$ in the present case. Using the well known properties of the Bernouilli shift $x \to Bx$, it is not difficult to show that the transform $P = C\circ B$ has a strange attractor made of horizontal segments $0 < x < 1$, $y = \frac{1}{8} \sum_{n=0}^{\infty} \frac{L_n}{4^n}$, where L_n is a set of numbers chosen at random as 1 or 5.

II.C - The boundaries of the attractor

This picture of the Poincaré transform is highly idealized, in particular the actual "longitudinal" transformation(i.e.the transformation law for x_2) is more complicated than the one of P. In order to make this point clearer, we have considered the intersect of the attractor by the half plane $x_1 = 35.0$ and $x_2 > x_1$ (owing to the general shape of the attractor its section by a given plane is made of two pieces at least, so that one has to define by some additional constraint, say $\dot{x}_1 > 0$, the piece in which one is interested). Let $x_2(n)$ be the value of x_2 when $x_1 = 35.0$ at the nth turn, we have plotted $x_2(n+1)$ as a function of $x_2(n)$ in Fig.8. This curve appears to be perfectly regular, at least at the accuracy of the computation. However it must have a fine structure as $x_2(n+1)$ does not depend on $x_2(n)$ only, but on another coordinate in the plane $x_1 = 35.0$, which would account for the finite thickness of the attractor and for the fact that the Poincaré transform must be invertible, this arises from the deterministic character of the system obtained from (1) by time inversion, although the 1d transform $x_2(n) \to x_2(n+1)$, as plotted on Fig.8 is obviously non invertible. But this may not represented in Fig.8, as the attractor looks like an ordinary surface in the 3d space.

Let us admit, for the moment that the Poincaré transform is a one dimensional transform. This allows one to account for two things : the attractor does not extend too far from the periodic trajectory, and when r reaches a critical value ($r \simeq 220$ approximately) the attractor changes to a stable limit cycle, without any qualitative change of the vector field.

It is clear from Fig.8 that only part of an apparently larger regular curve is reached and one may wonder why the values of $x_2(n)$ outside $[78.0, 98.8]$ do never occur. This may be understood as follows. Let $x_2 \to f(x_2)$ be the 1d Poincaré transform that we assume to be given on an interval larger than $[78.0, 98.8]$ by a curve that extents in a "natural" fashion the curve on Fig.8. Such an extension is drawn on Fig.9. (Note that the values of x_2 at the border of the attractor have been taken for simplicity as 0 and x_0 instead of 78.0 and 98.8). The system of axis in the plane $(x_2, f(x_2))$ is quite arbitrary, actually the quantities that resort from the equations of the motion are, in this plane, the bissectrix of the axis in the first quadrant and the curve itself. And the two border of the surface (or if one prefers, the points 0 and x_0) are determined by the condition that the segment $[0, x_0]$ applies on itself. Given the curve $f(x_2)$ and the bissectrix a simple geometrical construction yields these two points, as explained on (Fig.9). Furthermore two finite intervals $]-\varepsilon, 0[\ ]x_0, x_0+\varepsilon'[$, $\varepsilon, \varepsilon' > 0$, exists just outside the boundaries of $[0, x_0]$ such as, if x belongs to one of these intervals, then $f(x)$ either belongs to $[0, x_0]$

or at least is closer to x_o or 0 than x itself. If this condition of stability were not fullfilled, small fluctuations (as the one existing in analog or digital computations) occuring when the representative point is close to 0 or x_o, should drive the point far away from $[0,x_o]$, although one observes that the attractor is stable in this sense. The construction of these intervals is explained on (Fig.9). It is based upon the remark that the intersects of the curve f(x) with the bissectrix in the negative domain of x is an unstable fixed point. This limits the domain of attraction of $[0,x_o]$ on the left (i.e. the segment $]-\varepsilon,0]$) and, by a straightforward construction this is enough to draw the right part of the domain of attraction (i.e. $x_o,x_o+\varepsilon'$). As a conclusion of this discussion, let us notice that this does not mean that either 0 or x_o are fixed points or periodic points of the transforms.

By looking at this 1d Poincaré transform, one may understand the transition from stable limit cycle to strange attractor when r varies. In the present case all the fixed points are unstable, but a slight change in the Poincaré transform (that appears around $r \simeq 220$) makes the fixed point(or, eventually, periodic points) stable and the attractor becomes a periodic trajectory. Actually it has been shown[6)] for a one parameter quadratic transform which looks quite similar to $f(x_2)$ that a critical value of this parameter exists, say α_c, such as, if $\alpha < \alpha_c$ they are stable periodic points and if $\alpha > \alpha_c$ the transform has the property of mixing. In the present case, our investigations on the analog computer have shown that this dynamical system has the property of mixing, is that, given two smooth functions $\Phi(\vec{x})$ and $\psi(\vec{x})$ (actually, we have studied this property only for Φ and $\psi = x_i$ with $i = 1,2,3$, but it reasonable to assume that it remains true for any smooth Φ and ψ) then

$$\langle \phi[\vec{x}(t)]\ \psi\ [\vec{x}(t+T)] \rangle \xrightarrow[T\to]{} \langle\phi\rangle\langle\psi\rangle ,$$

where, by definition :

$$\langle \phi\ [\vec{x}(t)] \rangle = \lim_{\tau\to\infty} \frac{1}{\tau}\int_0^\tau dt\ \phi\left[\vec{x}(t)\right] .$$

This property means that, starting from two initial conditions very close to each other, then after some time the two representation points will be very far from each other, in other terms the systems "forgets" after some time its initial condition, which is perfectly compatible (this point is often a matter of misunderstandings) with the deterministic character of the equations of the motion.

III - A MODEL OF THE 2d POINCARE MAPPING

III.A - The construction of the Cremona transformation

As it is clear from the above considerations, one may consider the Poincaré mapping instead of the trajectories in the 3d spaces. This mapping is defined as a transformation of the plane S into itself : given a point A of S, we follow the trajectory which originates from A until it intersects S again. A trajectory is thus replaced by an infinite set of points in S, obtained by repeated application of the mapping T. The essential properties of the trajectory are reflected into corresponding properties of the set of points. We have thus formally reduced the problem to the study of a two-dimensional mapping.

At this point, however, the only advantage really gained is in clarity of presentation of the results ; the actual computation of the mapping still requires the numerical integration of the differential equations. Now comes the second and decisive step : we forget about the differential system, and we define a mapping T by explicit equations, giving directly T(A) when A is known. This of course simplifies the computation drastically. The new mapping T does not any more correspond to the Lorenz system ; however, by choosing it carefully we may hope to retain the essential properties which we wish to study. Past experience in the measure-preserving case (see reference 7 , and references therein) has shown indeed that the same features are found in dynamical systems defined by differential equations and in mappings defined as such.

The third step consists in specifying T . Here we have been inspired by the above explained numerical results on the Lorenz system, which show clearly how a volume is stretched in one direction, and at the same time folded over itself, in the course of one revolution. This folding effect has been also described by Ruelle (Ref. 4, Figs. 5 and 6). We simulate it by the following chain of three mappings of the (x, y) plane onto itself. Consider a region elongated along the x axis (Fig. 10a). We begin the folding by

$$T' : \quad x' = x \quad , \quad y' = y + 1 - a\,x^2 \quad , \tag{2}$$

which produces Fig. 10b ; a is an adjustable parameter. We complete the folding by a contraction along the x axis :

$$T'' : \quad x'' = b\,x' \quad , \quad y'' = y' \quad , \tag{3}$$

which produces Fig. 10c ; b is another parameter, which should be less than 1 in absolute value. Finally we come back to the orientation along the x axis by

$$T''' : \quad x''' = y'' \quad , \quad y''' = x'' \quad , \tag{4}$$

which results in Fig. 10d .

Our mapping will be defined as the product $T = T''' \, T'' \, T'$. We write now (x_i , y_i) for (x, y) and (x_{i+1} , y_{i+1}) for (x''', y''') (as a reminder that the mapping will be iterated) and we have

$$T : \quad x_{i+1} = y_i + 1 - a\,x_i^2 \quad , \quad y_{i+1} = b\,x_i \quad . \tag{5}$$

This mapping has some interesting properties. Its Jacobian is a constant :

$$\frac{\partial(x_{i+1} , y_{i+1})}{\partial(x_i , y_i)} = - b \quad . \tag{6}$$

The geometrical interpretation is quite simple : T' preserves areas ;

T''' also preserves areas but reverses the sign ; and T'' contracts areas, multiplying them by the constant factor b . The property (6) is welcome because it is the natural counterpart of the constant negative divergence in the Lorenz system.

A polynomial mapping satisfying (6) is known as an entire Cremona transformation, and the inverse mapping is also given by polynomials 8a,b). Indeed we have here

$$T^{-1} : \quad x_i = b^{-1} y_{i+1} \ , \quad y_i = x_{i+1} - 1 + a\, b^{-2} y_{i+1}^2 \ .$$

Thus T is a one-to-one mapping of the plane onto itself. This is also a welcome property, because it is the natural counterpart of the fact that in the Lorenz system there is a unique trajectory through any given point.

The selection of T could have been approached in a different way, by looking for the "simplest" non-trivial mapping. It is natural then to consider polynomial mappings of progressively increasing order. Linear mappings are trivial, so the polynomials must be at least of degree 2. The most general quadratic mapping is

$$x_{i+1} = f + a\, x_i + b\, y_i + c\, x_i^2 + d\, x_i\, y_i + e\, y_i^2 \ ,$$

$$y_{i+1} = f' + a'\, x_i + b'\, y_i + c'\, x_i^2 + d'\, x_i\, y_i + e'\, y_i^2 \qquad (7)$$

and depends on 12 parameters. But if we impose the condition that the Jacobian is a constant, some relations must be satisfied by these parameters. We can further reduce the number of parameters by an appropriate linear change of coordinates in the plane. In this way, by a slight extension of the results of Engel 8.b), it can be shown that the general form (7) is reducible to a "canonical form" depending on two parameters only. This is a generalization of the earlier result (reference 7) that a quadratic area-preserving mapping can be brought into a form depending on one parameter only. The canonical form can be written in several different ways ; and one of them turns out to be identical with (5), which is thus reached by an entirely different road ! The mapping (5), which was initially constructed in empirical fashion, is in fact the most general quadratic mapping with constant Jacobian.

One difference with the Lorenz problem is that the successive points obtained by repeated application of T do not always converge towards an attractor ; sometimes they "escape" to infinity. This is because the quadratic term in (5) dominates when the distance from the origin becomes large. However, for particular values of a and b it is still possible to prove the existence of a bounded "trapping region" R , from which the points can never escape once they have entered it (see below Section III.D).

T has two invariant points, given by

$$x = \frac{1}{2a}\left[-(1-b) \pm \sqrt{(1-b)^2 + 4a}\right] \quad , \quad y = b\,x \quad . \tag{8}$$

These points are real for

$$a > a_o = -\frac{1}{4}(1-b)^2 \quad . \tag{9}$$

When this is the case, one of the points is always linearly unstable, while the other is unstable for

$$a > a_1 = \frac{3}{4}(1-b)^2 \quad . \tag{10}$$

III.B - Choice of parameters

We select now particular values of a and b for a numerical study. b should be small enough for the folding described by Fig. 1 to occur really, yet not too small if one wishes to observe the fine structure of the attractor. The value $b = 0.3$ was found to be adequate. A good value of a was found only after some experimenting. For $a < a_0$ or $a > a_3$, where a_0 is given by (9) and a_3 is of the order of 1.55 for $b = 0.3$, the points always escape to infinity : apparently there exists no attractor in these cases. For $a_0 < a < a_3$, depending on the initial values (x_0 , y_0) , either the points escape to infinity or they converge towards an attractor, which appears to be unique for a given value of a . We concentrate now on this attractor. For $a_0 < a < a_1$, where a_1 is given by (10), the attractor is the stable invariant point. When a is increased over a_1 , at first the attractor is still simple and consists of a periodic set of p points. (An equivalent attractor in the Lorenz problem would be a limit cycle intersecting the surface of section p times). The value of p increases through successive "bifurcations" as a increases, and appears to tend to infinity as a approaches a critical value a_2 , of the order of 1.06 for $b = 0.3$. For $a_2 < a < a_3$, the attractor is no more simple, and the behaviour of the points becomes erratic. This is the case in which we are interested. We adopt the following values :

$$a = 1.4 \quad , \quad b = 0.3 \quad . \tag{11}$$

III.C - Numerical results

Fig. 11 shows the result of plotting 10 000 successive points, obtained by iteration of T, starting from the arbitrarily chosen initial point $x_0 = 0$, $y_0 = 0$; the vertical scale is enlarged to give a better picture. Fig.12 shows the result of 10 000 iterations of T again, starting from a different point : $x_0 = 0.63135448$, $y_0 = 0.18940634$ (this choice will be explained below). The two figures are seen to be almost identical. This suggests strongly that what we see in both figures is essentially the attractor itself : the successive points quickly approach the attractor and soon become undistinguishable from it at the scale of the figure. This is confirmed if one looks at the first few points on Fig.11. The initial point at $x_0 = 0$, $y_0 = 0$ and the first iterate at $x_1 = 1$, $y_1 = 0$ are clearly visible ; the second iterate is still visible at $x_2 = -0.4$, $y_2 = 0.3$; the third iterate can barely be distinguished at $x_3 = 1.076$, $y_3 = -0.12$; and the fourth iterate at $x_4 = -0.7408864$, $y_4 = 0.3228$ is already lost inside the attractor at the resolution of Fig. 2 . The following points then wander over the attractor in an apparently erratic manner.

One of the two unstable invariant points has the coordinates, given by (8) :

$$x = 0.63135448\ldots \ , \quad y = 0.18940634\ldots \tag{12}$$

This point appears to belong to the attractor. The two eigenvalues λ_1, λ_2 and the slopes p_1, p_2 of the corresponding eigenvectors are

$$\lambda_1 = 0.15594632\ldots \ , \quad p_1 = 1.92373886\ldots \ ,$$

$$\lambda_2 = -1.92373886\ldots \ , \quad p_2 = -0.15594632\ldots \ . \tag{13}$$

The instability is due to λ_2 . The corresponding slope p_2 appears to be tangent to the "curves" in Fig. 11.

These properties allow us to eliminate the "transient regime" in which the points approach the attractor, and which is not of much interest : we simply start from the close vicinity of the unstable point (12), by rounding off its coordinates to 8 digits. This is done in Fig.   and in the following figures. The points quickly move away along the line of slope p_2 since $|\lambda_2|$ is appreciably larger than 1.

The attractor appears to consist of a number of more or less parallel "curves" ; the points tend to distribute themselves densely over these curves. The few gaps that can still be seen on Figs. 11 and 12 have probably no particular significance. Their locations are not the same on the two figures. They are simply due to statistical fluctuations in the quasi-random distribution of points, and they would disappear if more moints were plotted. Thus, the <u>longitudinal structure</u> of the attractor (along the curves) appears to be simple, each curve being essentially a one-dimensional manifold.

The <u>transversal structure</u> (across the curves) appears to be entirely different, and much more complex. Already on Figs. 11 and 12 a number of curves can be seen, and the visible thickness of some of them suggests that they have in fact an underlying structure. Fig. 13 is a magnified view of the small square of Fig. 12 : some of the previous "curves" are indeed resolved now into two or more components. The number n of iterations has been increased to 10^5 , in order to have a sufficient number of points in the small region examined. The small square in Fig. 13 is again magnified to produce Fig. 14 , with n increased to 10^6 : again the number of visible "curves" increases. One more enlargement results in Fig. 15 , with $n = 5 \times 10^6$: the points become sparse but new curves can still easily be traced.

These figures strongly suggest that the process of multiplication of "curves" will continue indefinitely, and that each apparent "curve" is in fact made of an infinity of quasi-parallel curves. Moreover, Figs. 13 to 14 indicate the existence of a hierarchical sequence of "levels", the structure being practically identical at each level save for a scale factor. This is exactly the structure of a Cantor set.

The frames of Figs. 13 to 14 have been chosen so as to contain the invariant point (12) . This point appears to lie on the upper boundary of the attractor. Surprisingly, its presence is completely invisible on the figures ; this contrasts with the area-preserving case, where stable and unstable invariant points play a very conspicuous role[7)] . On the other hand, the presence of the invariant point explains, locally at least, the hierarchy of similar structures : at each application of the mapping, the scale of the transversal structure is multiplied by λ_1 given by (13) . At the same time, the points spread out along the curves, as dictated by the value of λ_2 .

III.D - A trapping region

The fact that even after 5×10^6 iterations the points have not diverged to infinity suggests that there is a region of the plane from which the points cannot escape. This can be actually proved by finding a region R which is mapped inside itself. An example of such a region is the quadrilateral ABCD defined by

$$x_A = -1.33 \ , \quad y_A = 0.42 \ , \quad x_B = 1.32 \ , \quad y_B = 0.133 \ ,$$

$$x_C = 1.245 \ , \quad y_C = -0.14 \ , \quad x_D = -1.06 \ , \quad y_D = -0.5 \ . \qquad (14)$$

The image of ABCD is a region bounded by four arcs of parabola, and it can be shown by elementary algebra that this image lies inside ABCD . Plotting the quadrilateral on Fig. 11 or 12, one can verify that it encloses the observed attractor.

III.E - Conclusions

The simple mapping (5) appears to have the same basic properties as the Lorenz system. Its numerical exploration is much simpler : in fact most of the exploratory work for the present paper was carried out with a programmable pocket computer (HP-65). For the more extensive computations of Figs. 11 to 14, we used a IBM 7040 computer, with 16-digit accuracy. The solutions can be followed over a much longer time than in the case of a system of differential equations. The accuracy is also increased since there are no integration errors.

We inferred in first part the Cantor set structure from reasoning, but oould not observe it directly because the contracting ratio after one 3circuit3 was too small : 7×10^{-5}. In the present mapping, the contracting ratio after one iteration is 0.3, and one can easily observe a number of successive levels in the hierarchy. This is also facilitated by the larger number of points.

Finally, for mathematical studies the mapping (5) might also be easier to handle than a system of differential equations.

REFERENCES

1) E.N. Lorenz, J. Atmo Sciences 20, 130 (1963)

2) R. Thom, "Stabilité Structurelle et morphogénèse : Essai d'une théorie générale générale des modèles", Reading, Massachussets : Addison Wesley (1973)

3) T. Rikitake, Proc. Cambridge Phil. Soc. 54, 89 (1958)
D.W. Allan, ibid. 38, 671 (1962)

J.H. Mathews and W.K. Gardner, Naval res. Lap. Resp. 5886 (1963), Kay Ann Robbins, PhD. Thesis (unpublished) Massachussetts Institute of Technology (June 1975).

4) D. Ruelle, report at the conference on "Quantum dynamics models and mathematics" Bielefeld, September 1975, O. Landford III, unpublished word

5) G.D. Birkhoff, Trans. Amer. Math. Soc. 18, 199 (1917)

6) T.Y. Li and J. Yorke, preprint
R.H. May, Science 186, 645 (1975)

7) M. Hénon, Quaterl. Appl. Math. 27, 299 (1969)

8a) W. Engel, Math. Annal. 130, 11 (1955), and (8b) 136, 319 (1958)

FIGURE CAPTIONS

Fig.1

Projections of the stable limit cycle found at $r = 230$.

Fig.2

Projections of the strange attractor found at $r = 210$. This attractor is drawn by trajectories which fluctuate around a periodic (unstable) cycle quite similar to the one of Fig.1. The two fixed points $(+\sqrt{\beta(r-1)}, \pm\sqrt{\beta(r-1)}, r-1)$ are indicated by two crosses.

Fig.3

Projections of the sections of the attractor by series of planes parallel to x_1.

Fig.4

The unstable periodic trajectory at $r = 210$. The sections normal to the attractor that are drawn on Fig.5 have been taken at points indicated by A,B,C... on this curve. The arrow indicates the direction of the motion.

Fig.5A and 5B

These are the sections of the attractor by planes normal to the periodic trajectory at points A,B,C,.... The number in parenthesis indicate the magnification of the sections, which is needed for getting drawings of similar size. The circled point is on the periodic trajectory.

Fig.6A,6C,6D,6E,6G and 6K

A few examples of sections of the attractor by planes normal to the mean trajectory together with the component of the velocity field in this plane. The drawing numbered 6A corresponds to section A, the one numbered 6C to section C, and so on.

Fig.7

On the left hand side a shematic representation of the Poincaré transform is given. The drawing of the first iterates illustrate

Fig.7 (cont'nd)

the idea that a single sheet survive after an infinite number of steps without being cut anywhere. The right side illustrates more exactly the successive steps of the transform, and account for the asymmetry of the folding process.

Fig.8

This graph is an experimental plot of $x_2(n+1)$ as a function of $x_2(n)$ in the plane $x_1 = 35.0$.

Fig.9

This explains the construction of the border of the attractor, the 1d Poincaré transform being given. One draws first the curve symmetrical with respect to the bissectrix of the axis (i.e. the line $x_2(n) = x_2(n+1)$) then the straight tangent to the curve at its maximum. This tangent intersects the symmetric of the curve $x_2(n+1) = f(x_2(n))$ at a point C and the straight vertical drawn from C defines the left limit of the attractor. The construction of the border on the right is straightforward.

Fig.10

The initial area (a) is mapped by T' into (b), then by T" into (c), and finally by T" into (d).

Fig.11

10 000 successive points obtained by iteration of the mapping T starting from $x_o = 0$, $y_o = 0$.

Fig.12

Same as Fig. 11, but starting from $x_o = 0.63135448$, $y_o = 0.18940634$.

Fig.13

Enlargement of the squared region of Fig. 12. The number of computed points is increased to $n = 10^5$.

Fig.14

Enlargement of the squared region of Fig. 13; $n = 10^6$.

Fig.15

Enlargement of the squared region of Fig. 14; $n = 5 \times 10^6$.

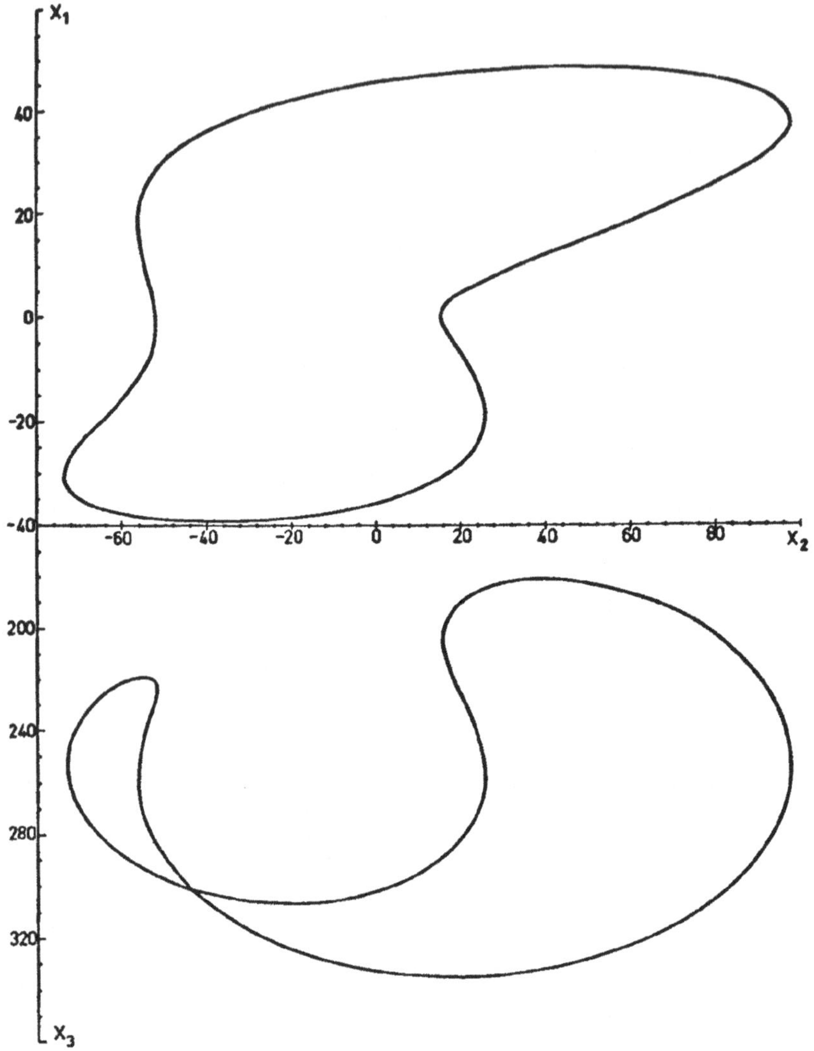

Fig. 1

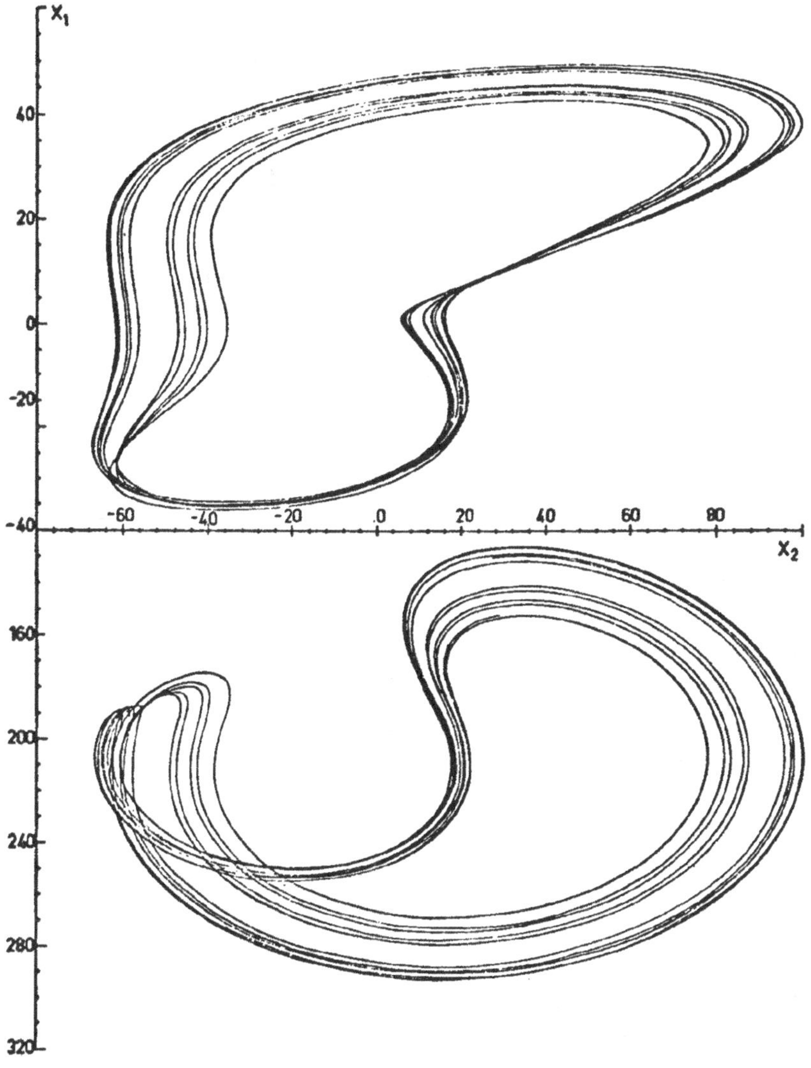

Fig. 2

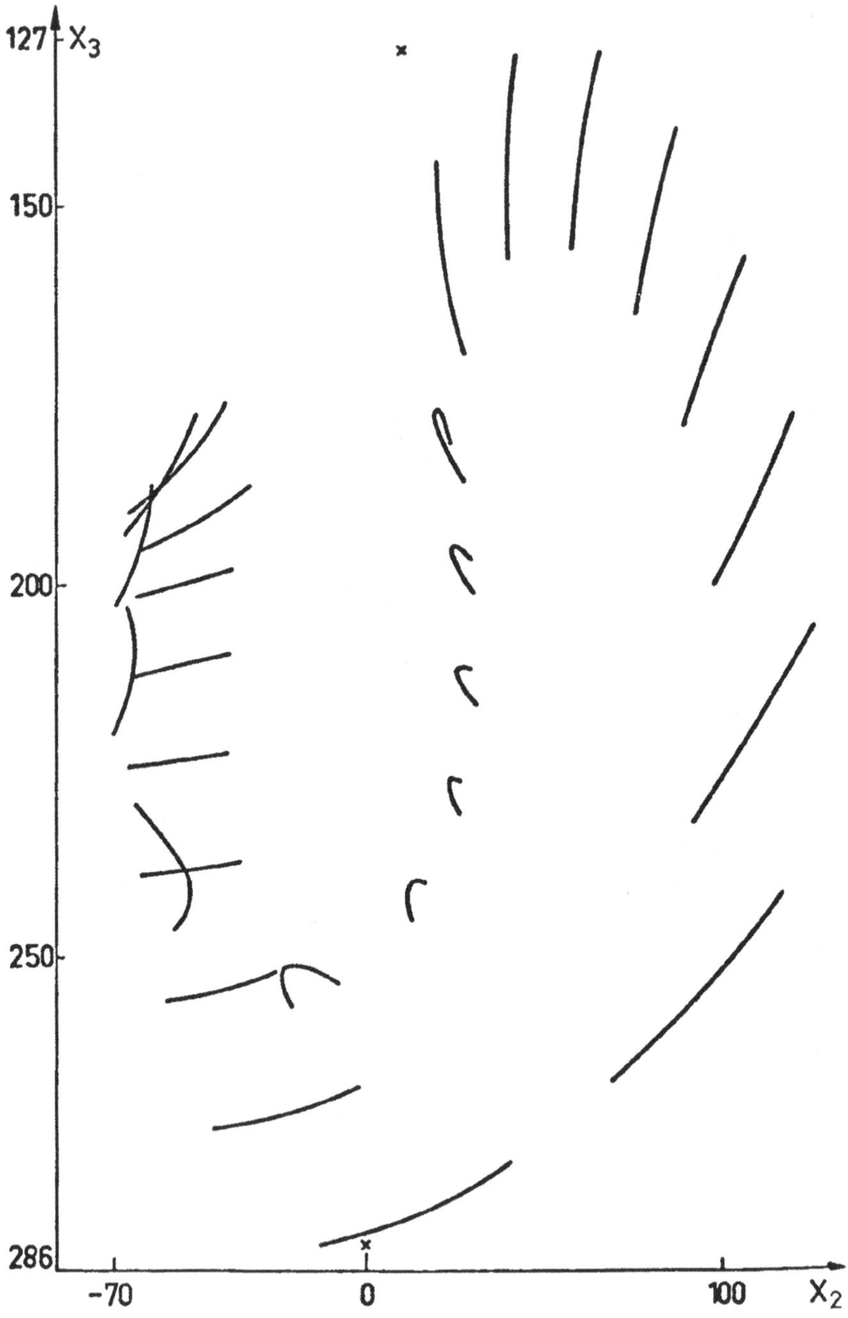

127
X_3
150
200
250
286
-70
0
100
X_2

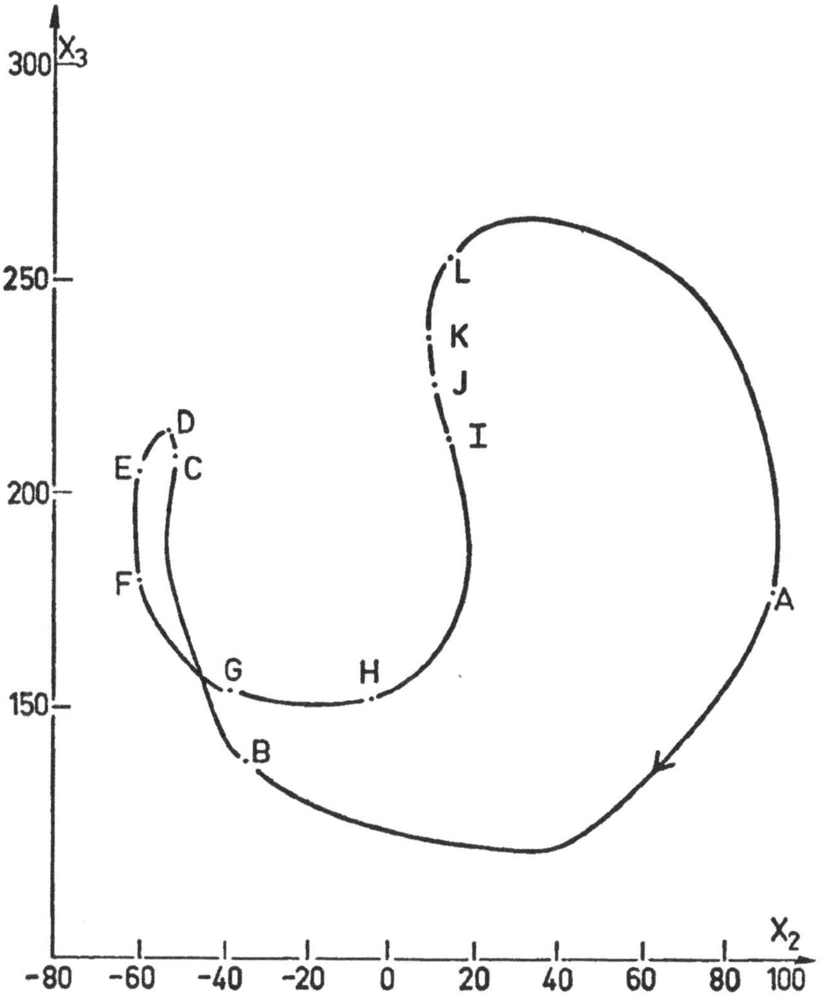

Fig. 4

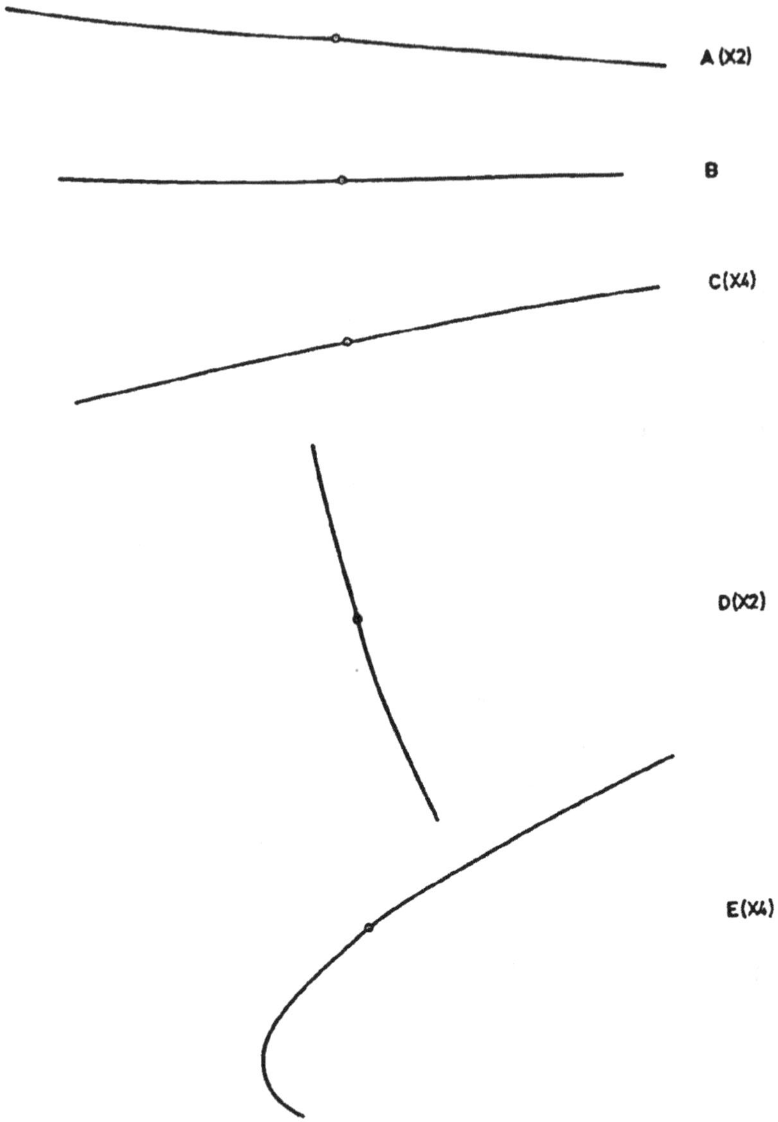

Fig. 5 A

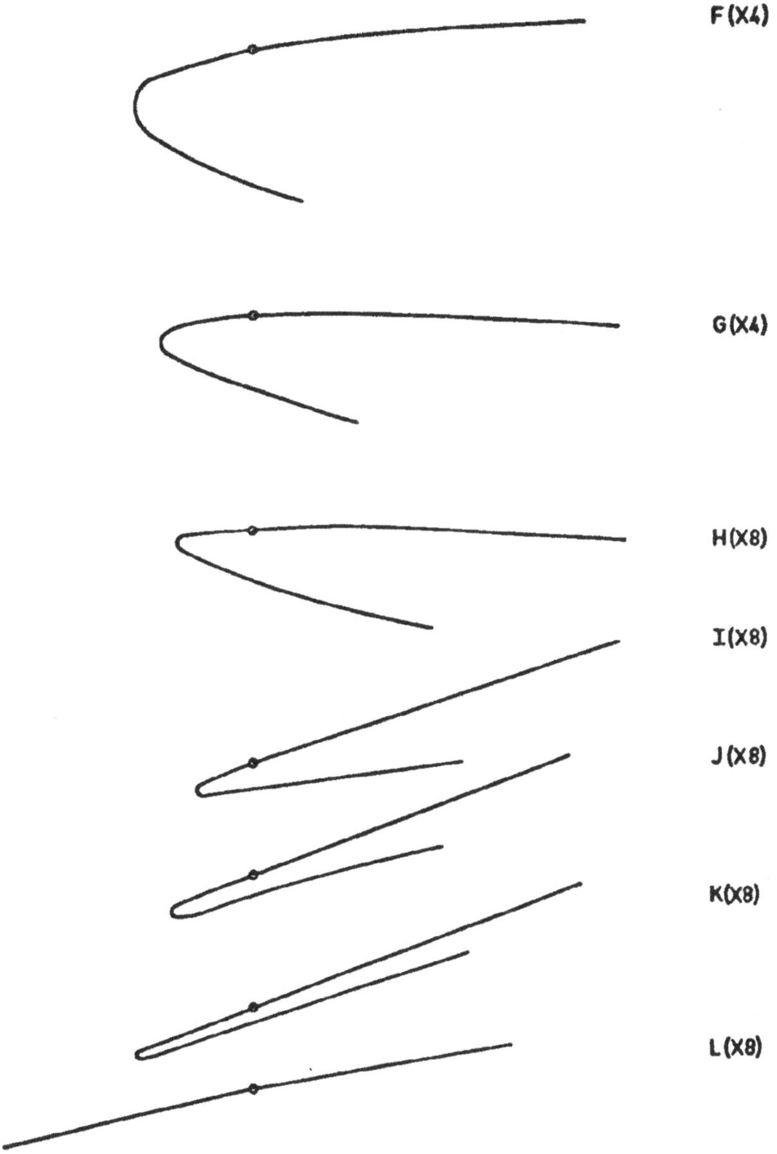
F(X4)
G(X4)
H(X8)
I(X8)
J(X8)
K(X8)
L(X8)

Fig. 5 B

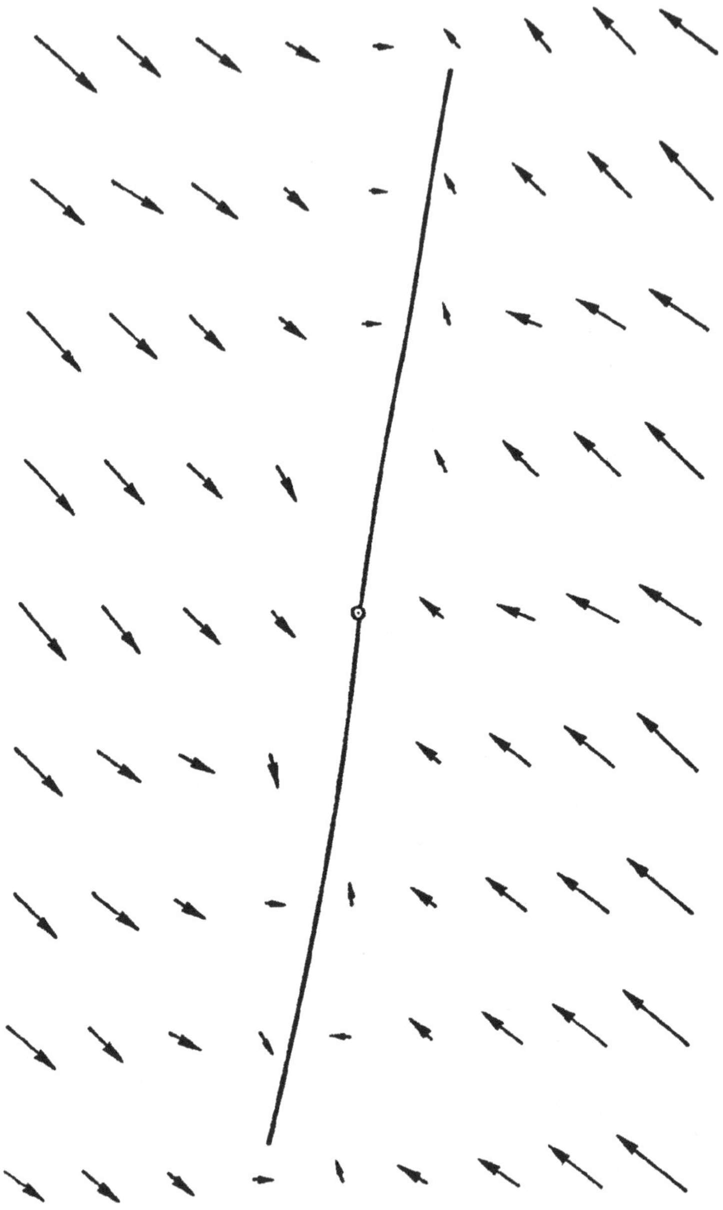

Fig. 6 A

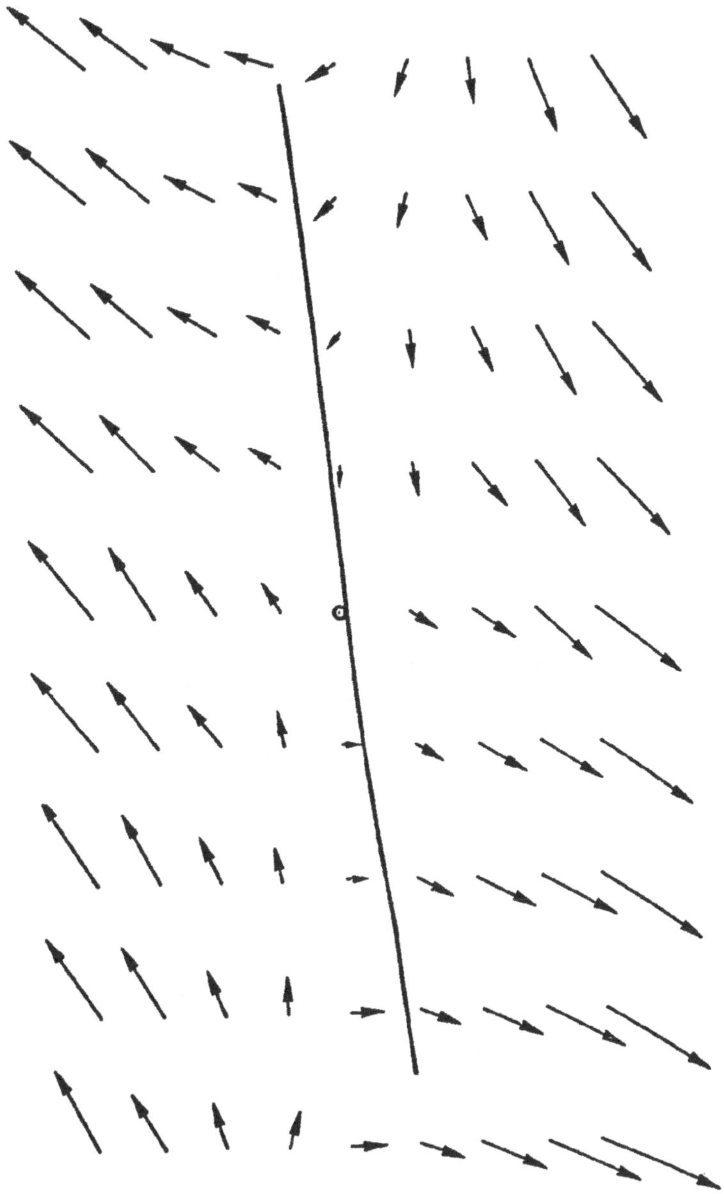

Fig. 6 C

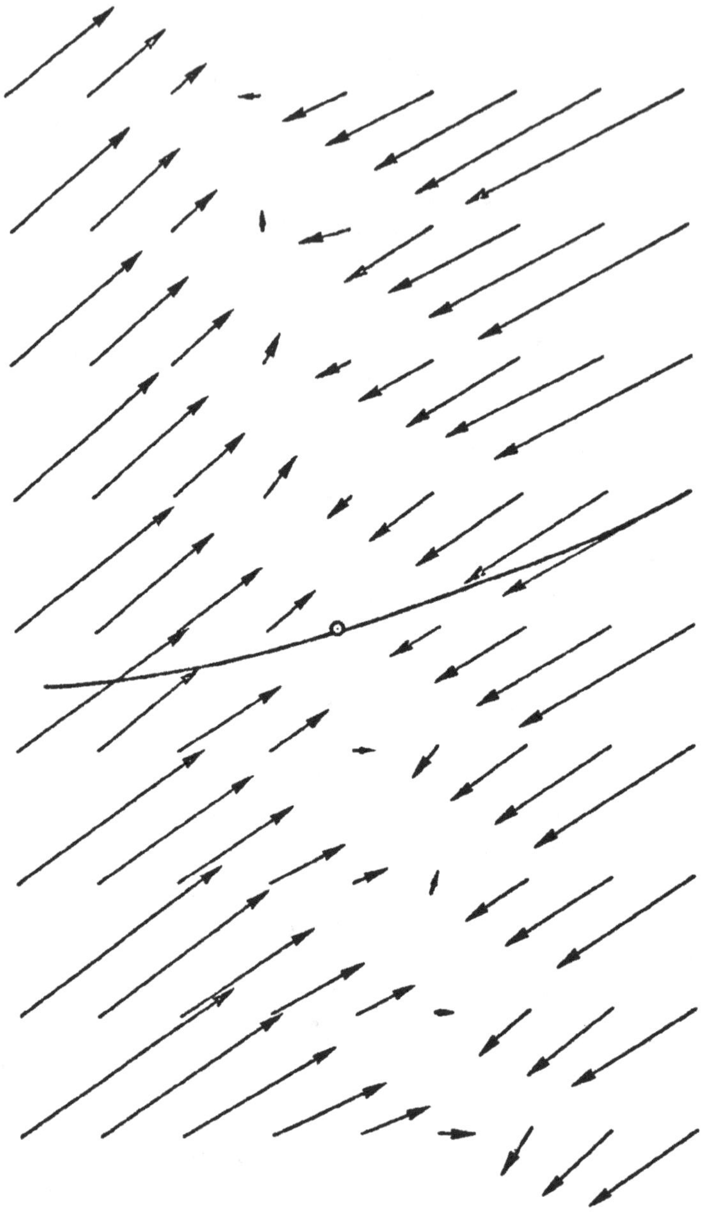

Fig. 6 D

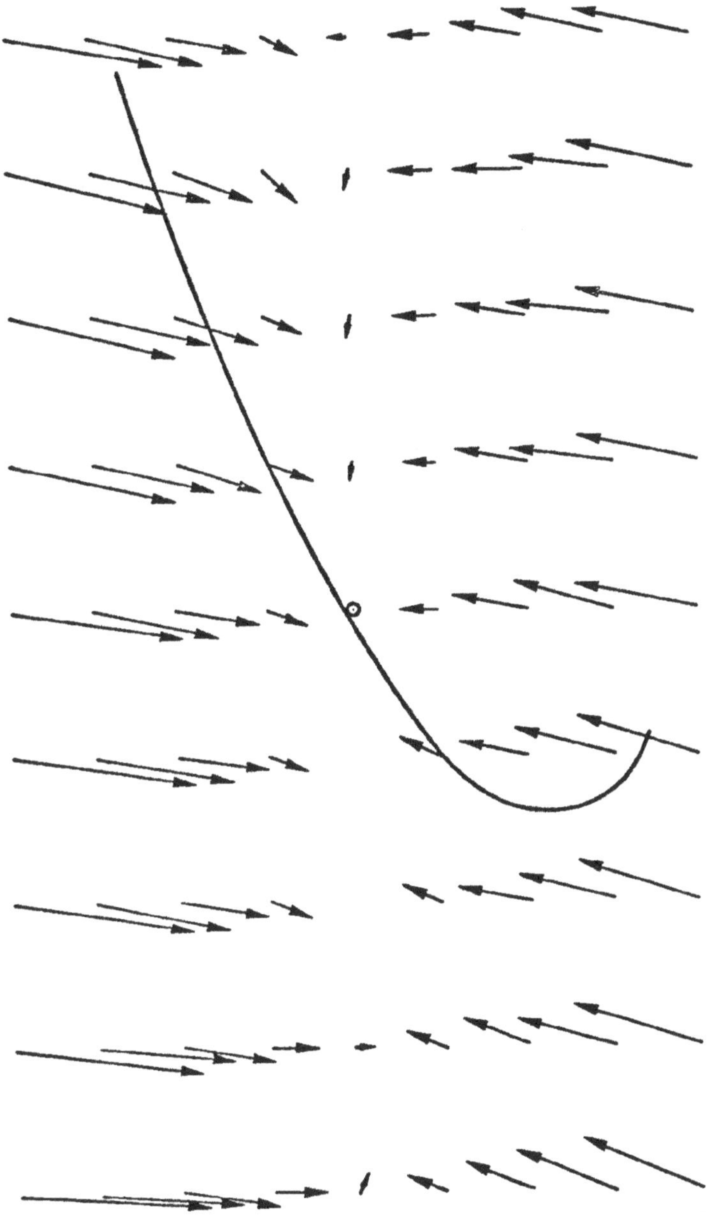

Fig. 6 E

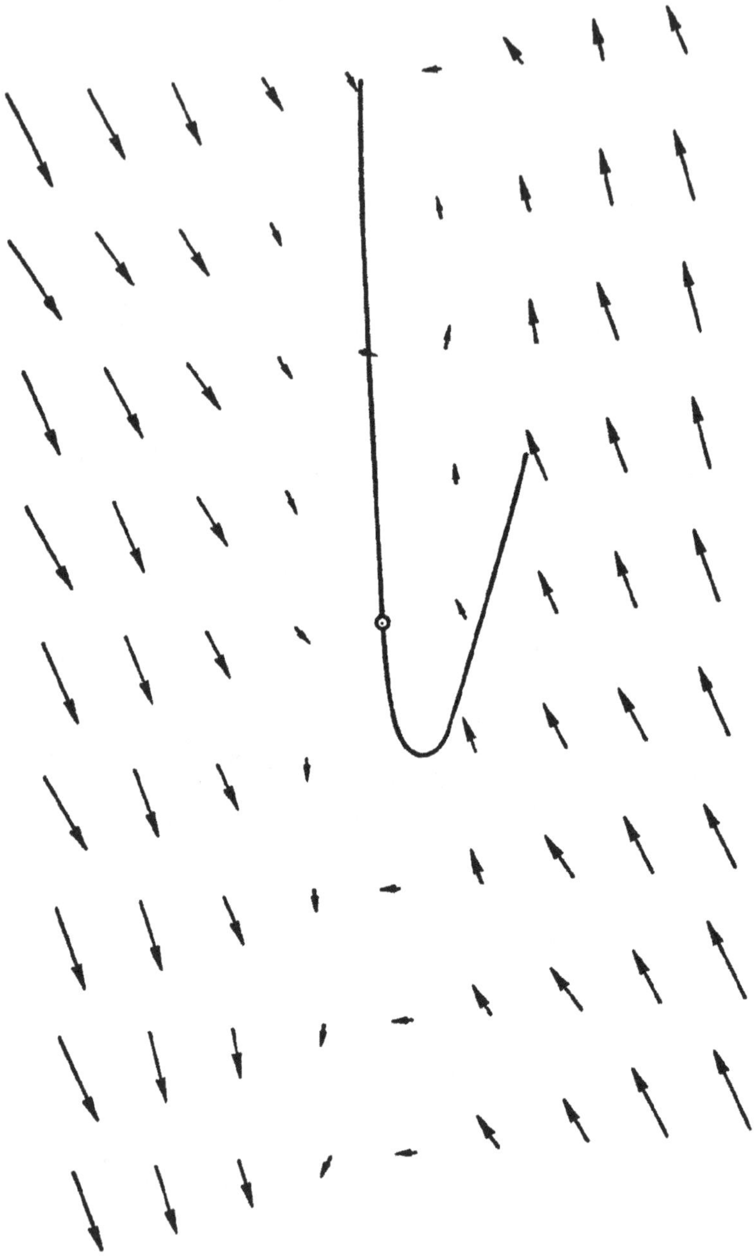

Fig. 6 G

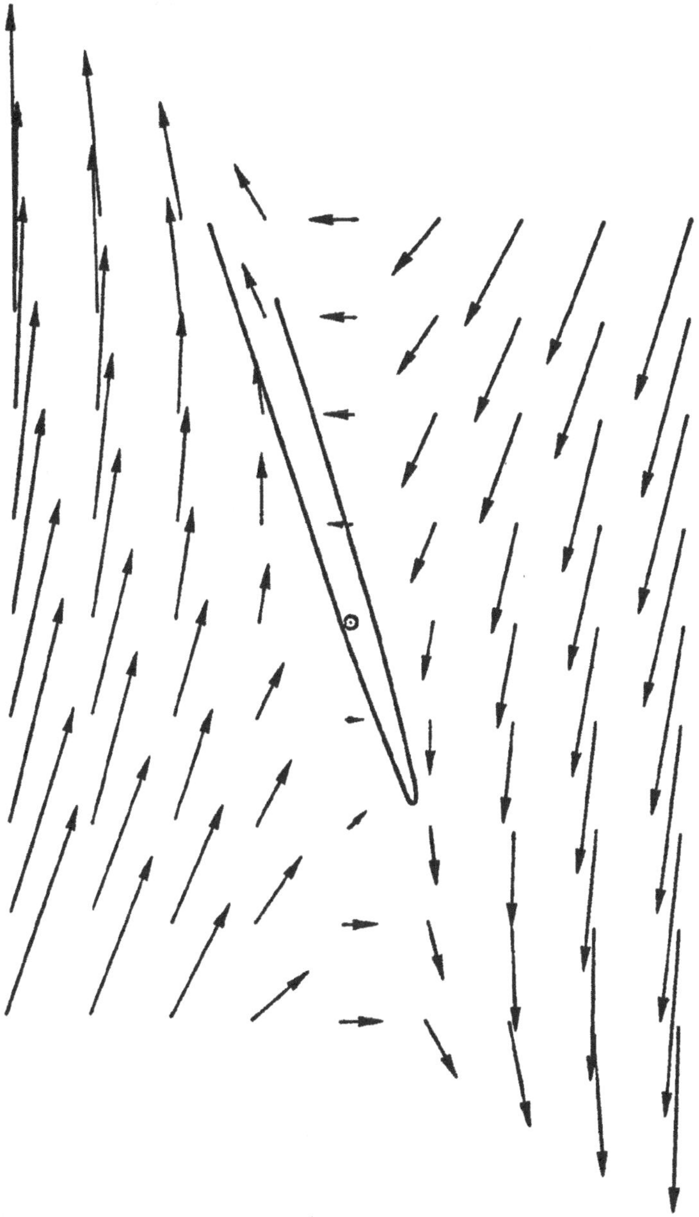

Fig. 6 K

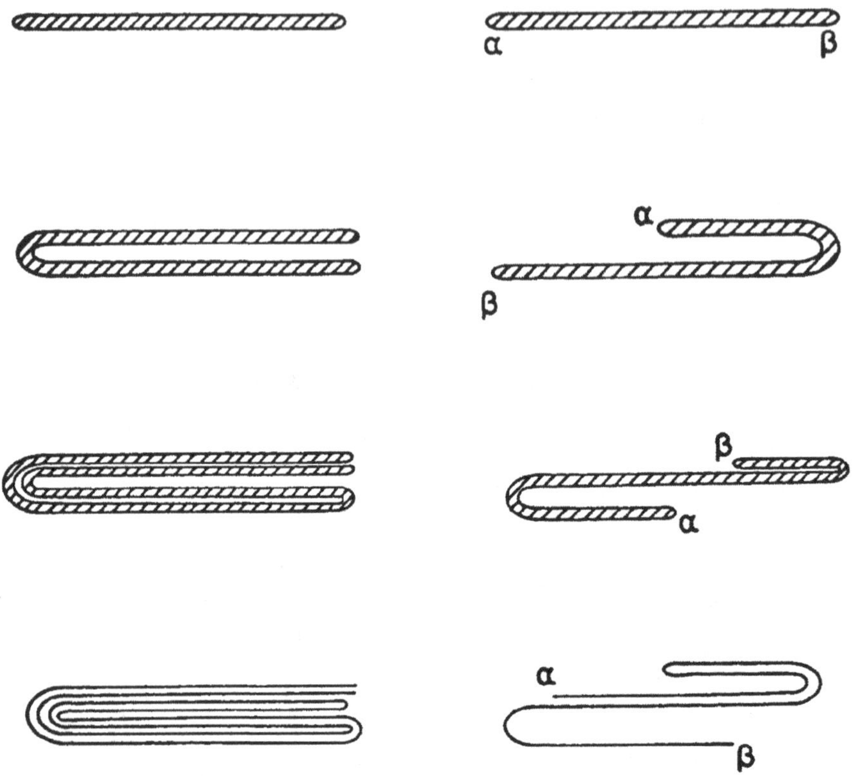

Fig. 7

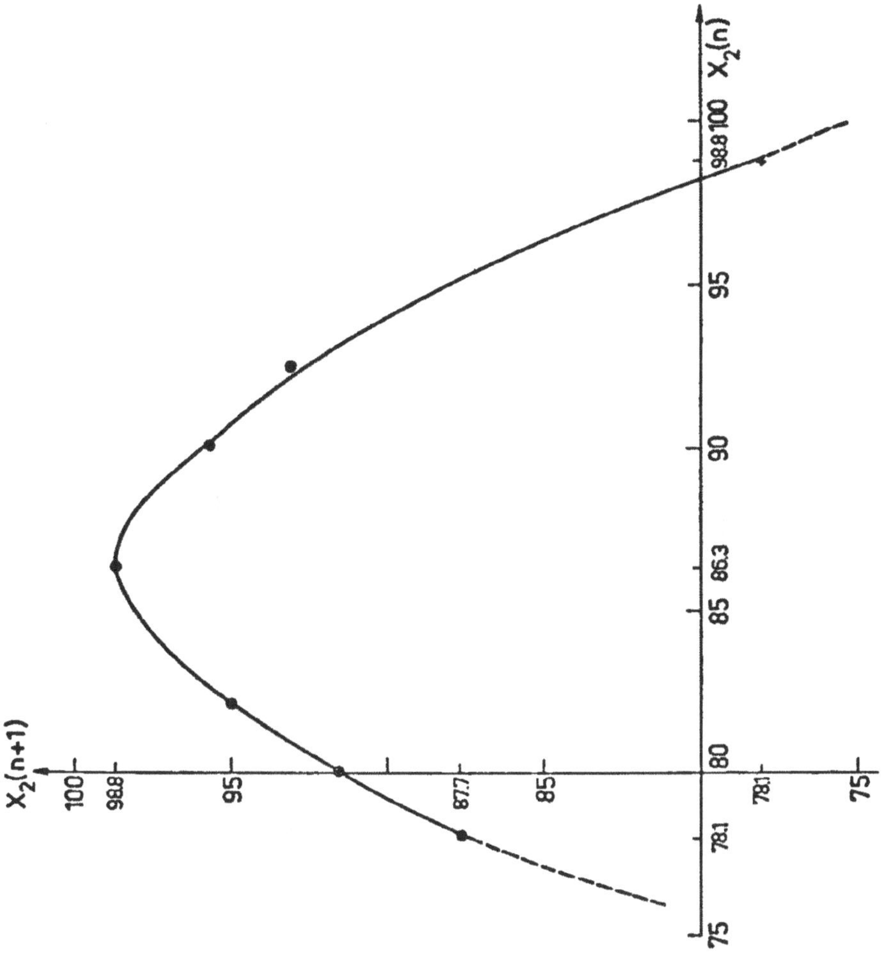

Fig. 8

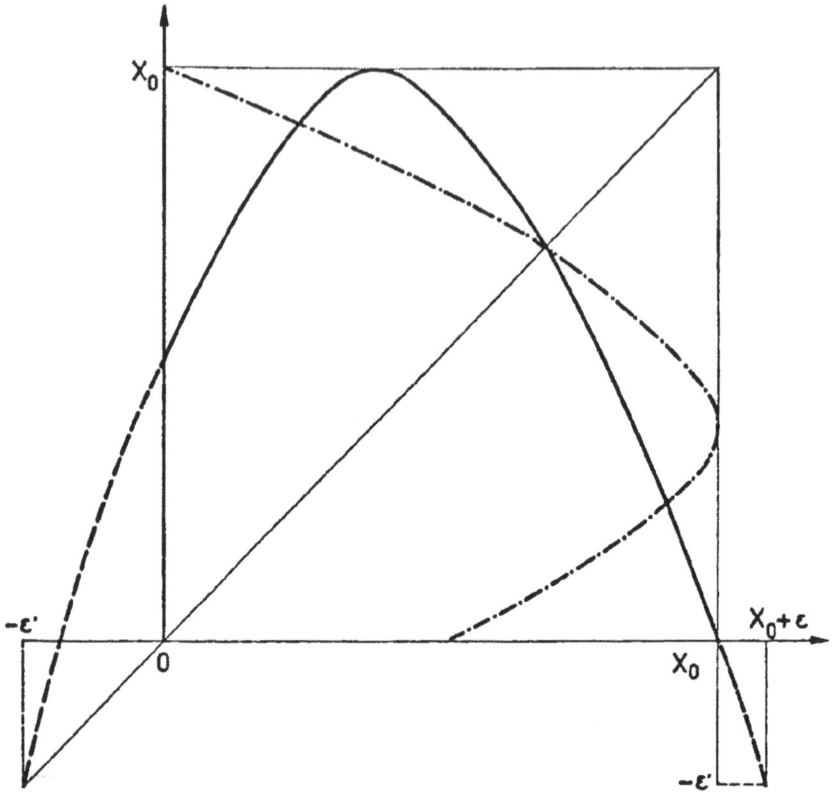

Fig. 9

Fig. 10

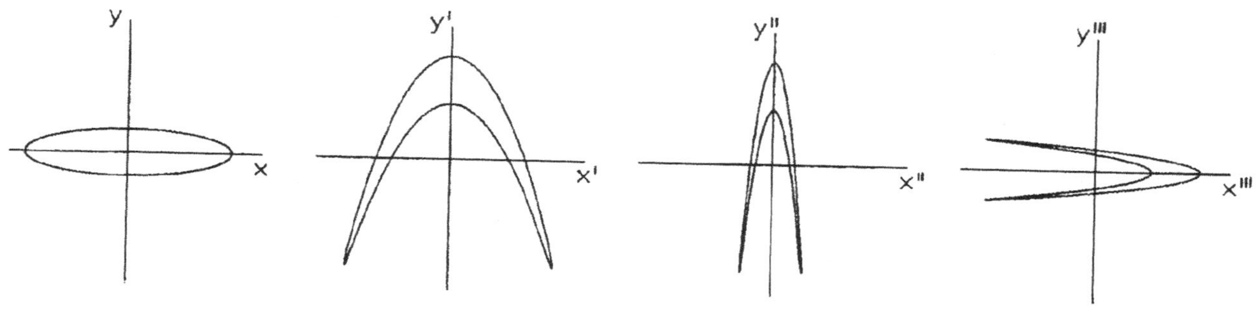

a b c d

Fig. 11

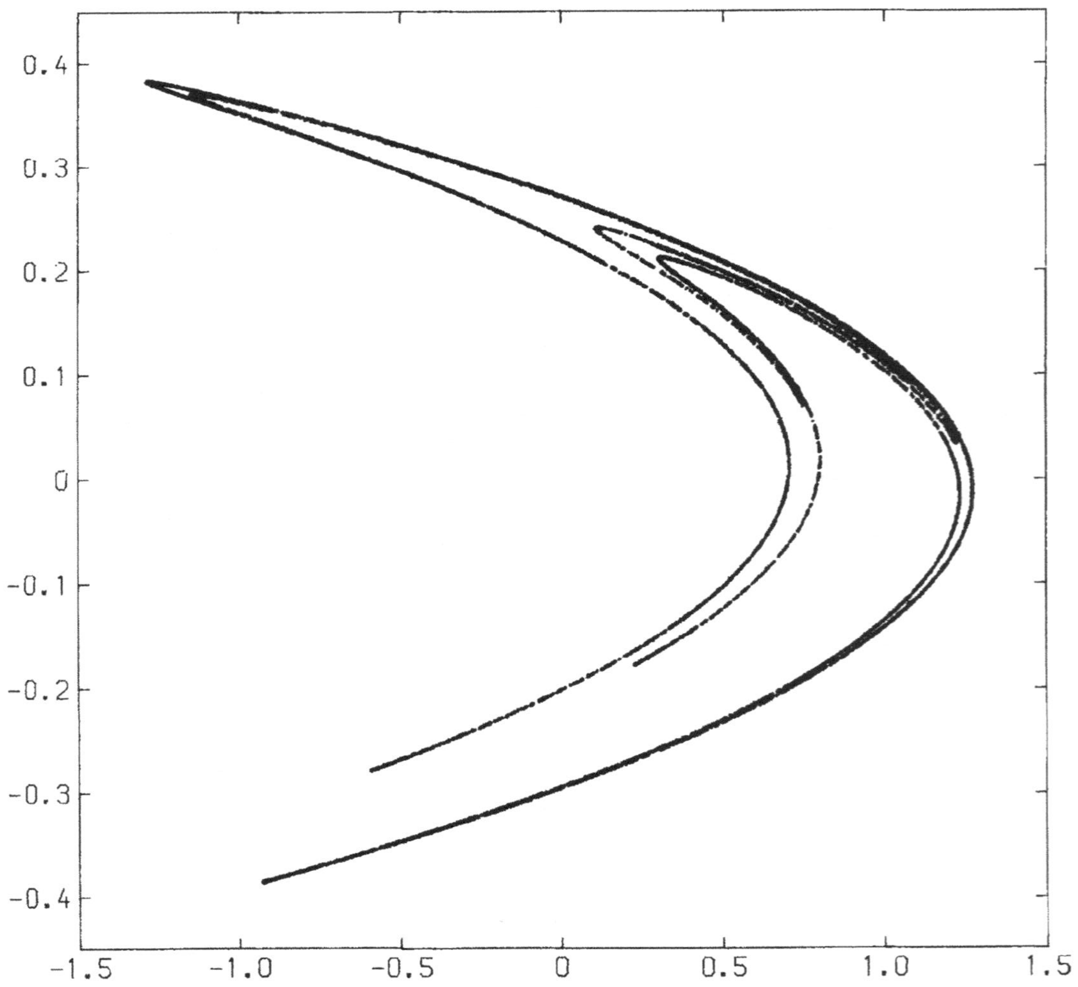

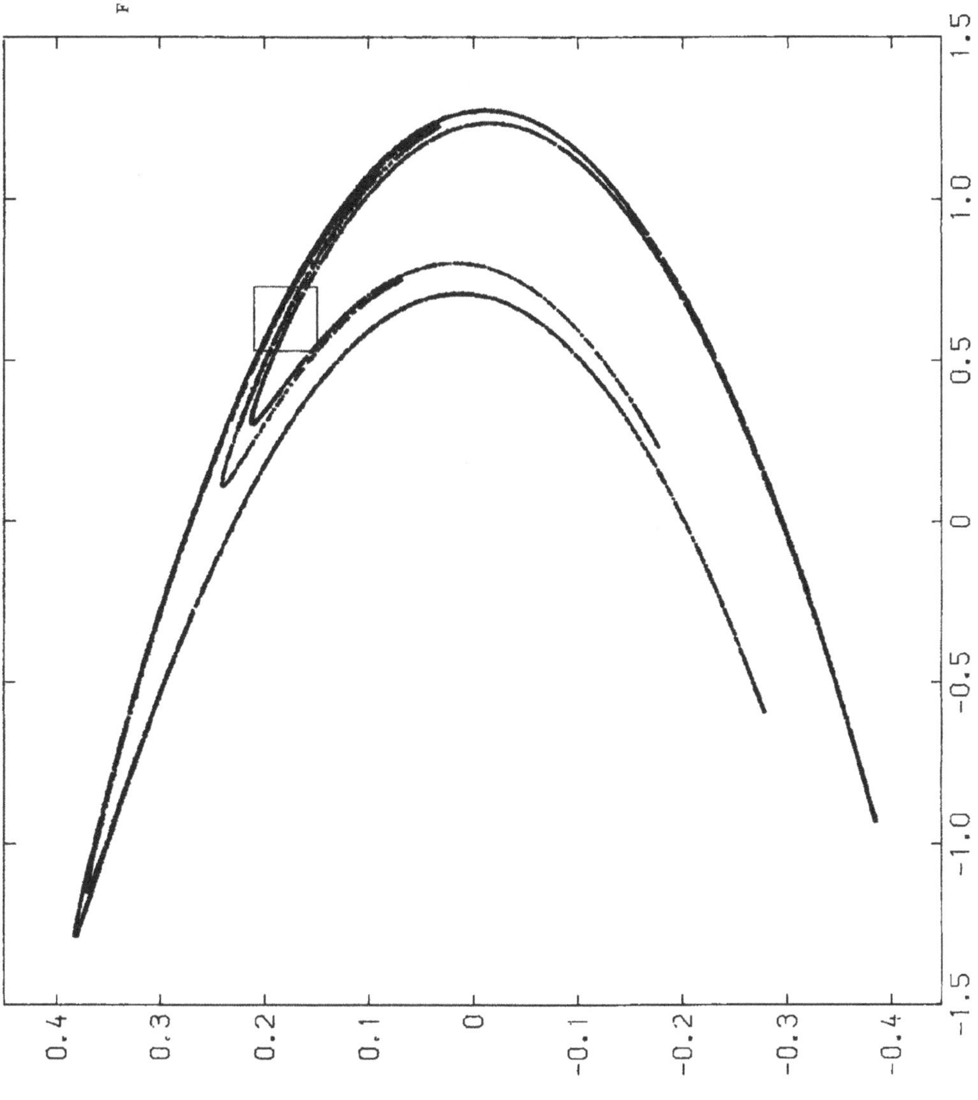

Fig. 12

Fig. 13

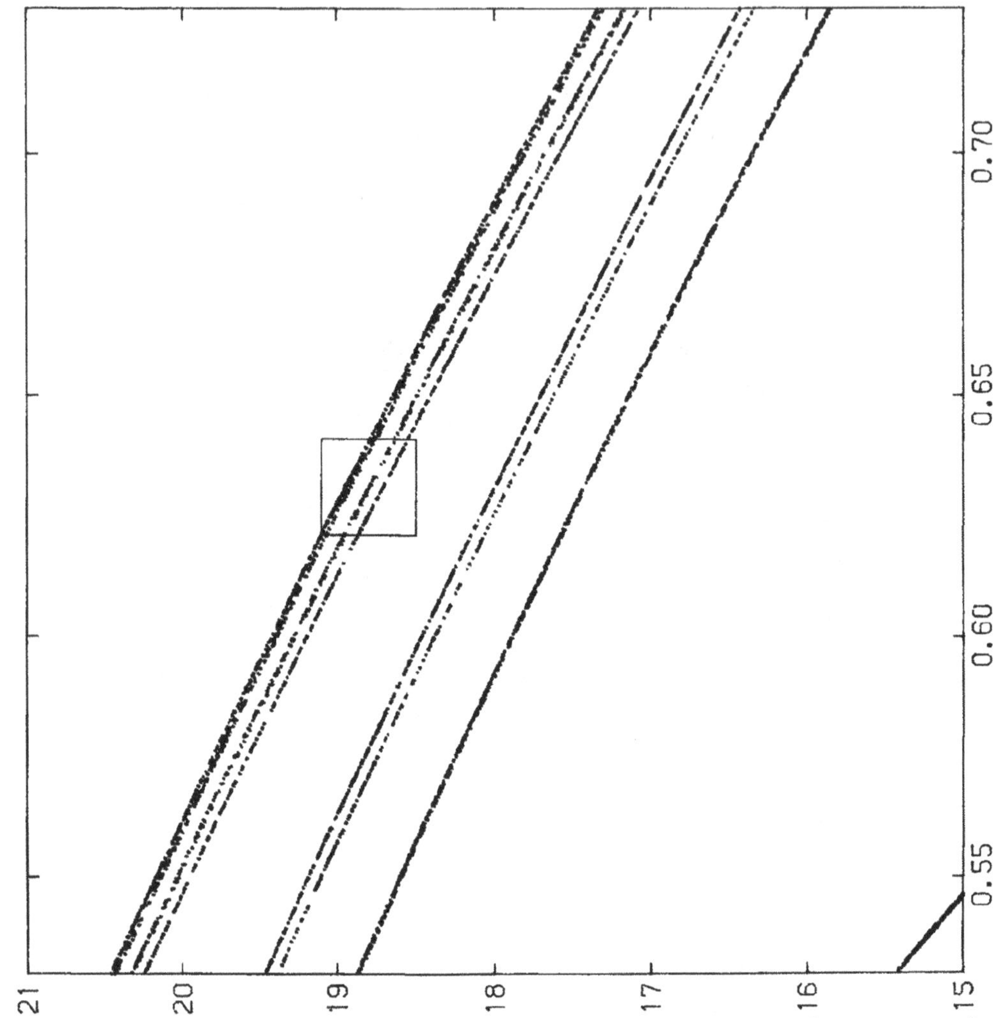

Fig. 14

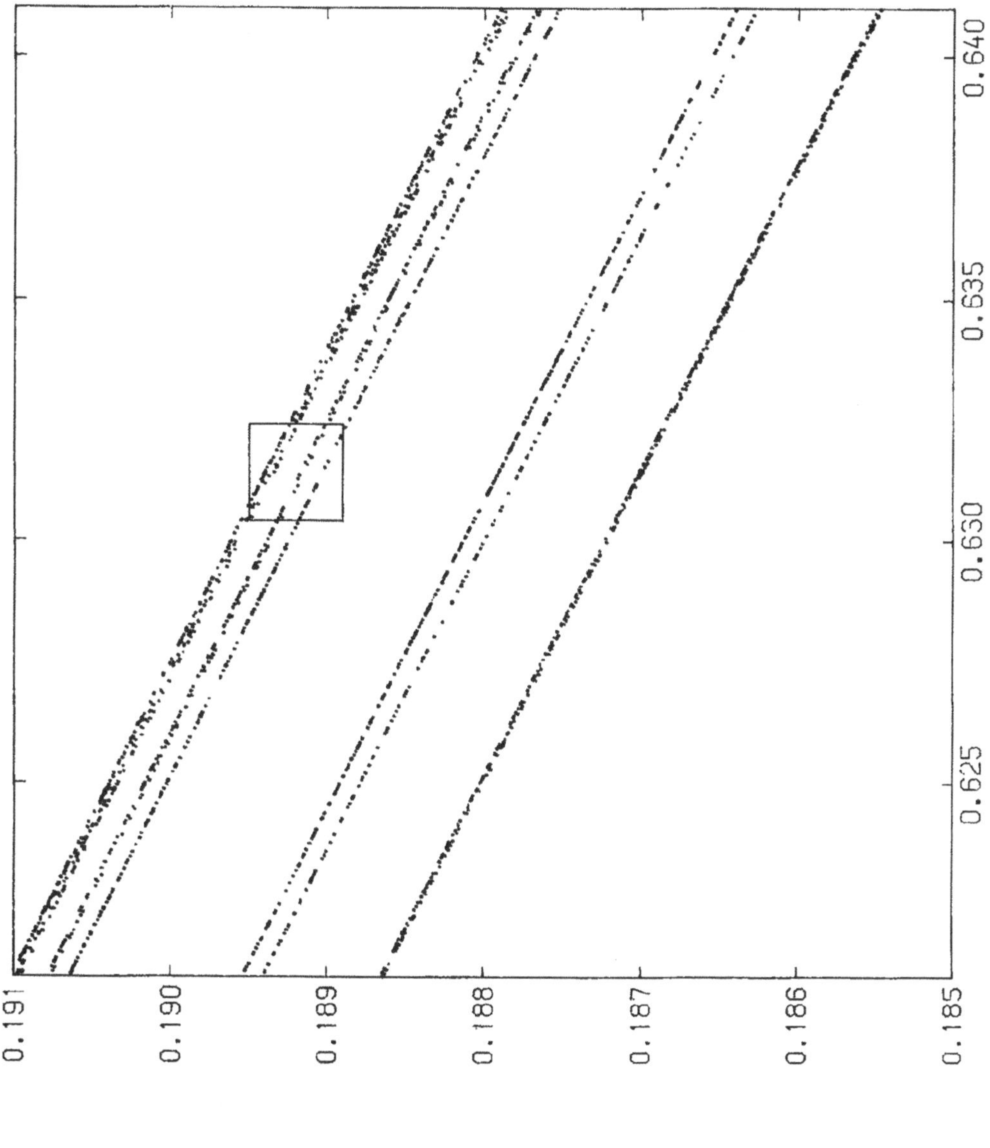

Fig. 15

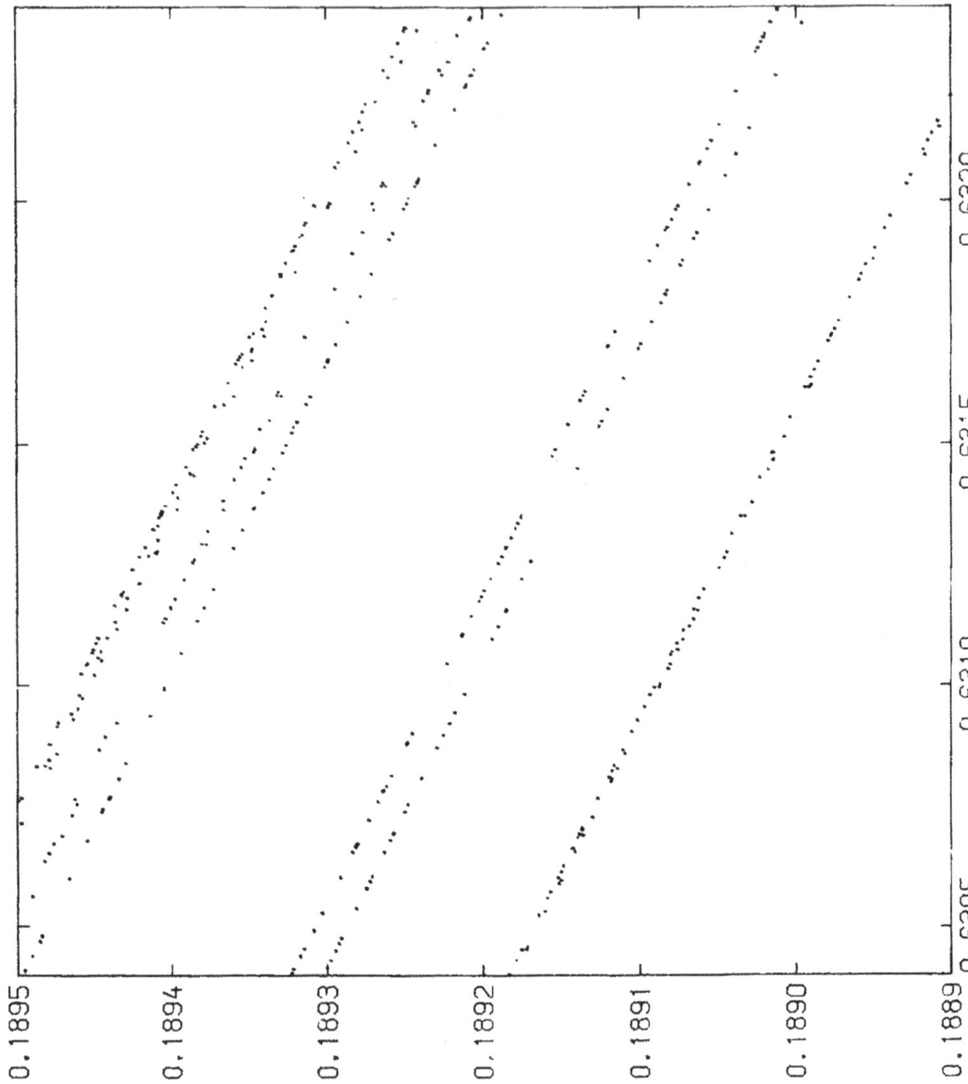
0.6305
0.6310
0.6315
0.6320
0.1895
0.1894
0.1893
0.1892
0.1891
0.1890
0.1889

DIRECT BIFURCATION OF A STEADY SOLUTION OF THE NAVIER-STOKES EQUATIONS INTO AN INVARIANT TORUS

Gérard IOOSS
Institut de Mathématiques et Sciences Physiques
Parc Valrose, 06034 NICE Cedex (France)

Foreword The communication presented at the "Journées Mathématiques sur la Turbulence" was divided in two parts. The first part treated the secondary bifurcation of a steady solution into an invariant torus ; this is writen in a paper [5]. All details of proofs on this first part will appear in [6] . Now, we give here all details on the second part of the communication.

I - Statement of the problem

1. Let us consider a viscous incompressible flow in a bounded regular domain $\Omega \subset \mathbf{R}^3$ or $\mathbf{R}^2$, satisfying the Navier-Stokes equations :

$$(1) \qquad \begin{cases} \left.\begin{array}{l} \dfrac{\partial v}{\partial t} + (v.\nabla)v + \nabla p = \nu \Delta v + f \\ \nabla . v = o \end{array}\right\} \text{ in } \Omega \\ v\big|_{\partial\Omega} = a \quad , \text{ where } \displaystyle\int_{\partial\Omega} a.n\, ds = o \end{cases}$$

v is the velocity of the fluid at the point $(x,t) \in \Omega \times \mathbf{R}_+$, p is the pressure, f is a given external force, a is given on the boundary $\partial\Omega$, and ν is the reciprocal of the Reynolds number. In all the following, the system (1) is considered as an example of a system sitting in our frame. In fact, we can consider such systems as those which occur in Bénard-convection, or in magnetohydrodynamic flows.

2. Let us assume that we know a steady solution (v_o, p_o) of (1). This solution is called "the basic flow". Now, following the problem we consider, we have a characteristic parameter, such as ν^{-1} or any parameter occuring in f or a. Let us denote it by λ , and assume that v_o is analytic in λ. Now we pose

$$v = v_o(\lambda) + u .$$

Hence, the perturbation u satisfies a system of the form

$$(2) \qquad \frac{du}{dt} + L_\lambda u - M(u) = o ,$$

where we look for $t \mapsto u(t)$ as a continuous function taking values in the domain of the linear operator L_λ , with a continuous derivative in an Hilbert space H . In the case of the system (1) , we introduce the following Hilbert spaces (with standard scalar products) :

$$H = \{u \in [L^2(\Omega)]^3 \; ; \; \nabla . u = o \quad , \; u . n|_{\partial\Omega} = o\} \quad ,$$
$$K = \{u \in [H^1(\Omega)]^3 \; ; \; \nabla . u = o \quad , \; u . n|_{\partial\Omega} = o\} \quad ,$$
$$\mathcal{D} = \{u \in [H^2(\Omega)]^3 \; ; \; \nabla . u = o \quad , \; u|_{\partial\Omega} = o\} .$$

Let us denote by Π the orthogonal projection in $[L^2(\Omega)]^3$ onto H .
It is known ([12] , [16]) that $H^\perp = \{u = \nabla\varphi \; ; \; \varphi \in H^1(\Omega)\}$ and that $\Pi \in \mathcal{L}([H^1(\Omega)]^3 ; K)$ where $\mathcal{L}(\mathfrak{N}_1 ; \mathfrak{N}_2)$ denotes the Banach space of bounded linear operators from $\mathfrak{N}_1$ into $\mathfrak{N}_2$. Now, we have $\forall u \in \mathcal{D}$

$$L_\lambda u = \Pi[-\nu \Delta u + (u . \nabla) v_o(\lambda) + (v_o(\lambda) . \nabla) u] \in H ,$$
$$M(u) = -\Pi[(u . \nabla) u] \in K .$$

3. Let us enumerate the properties of the operators in (2) , in a general form, in the aim to be applied to other systems as (1) .

i) We have the continuous imbeddings :

$$\mathcal{D} \hookrightarrow K \hookrightarrow H$$

for the 3 Hilbert spaces H, K, $\mathcal{D}$, where the norm in $\mathcal{D}$ is constructed with the norm of the graph of L_{λ_o} .

ii) $\{L_\lambda\}_{\lambda \in D_o}$ is an holomorphic family of closed operators in H , with domain $\mathcal{D}$, where D_o is a domain of $\mathbb{C}$. We assume that L_λ has a compact resolvant, and that this operator is real if $\lambda \in \mathbb{R}$.

iii) $\forall \lambda \in D_o$, it can be defined an holomorphic semi-group of operators in H : $\{e^{-L_\lambda t}\}_{t \geq o}$. We assume that $t \mapsto e^{-L_\lambda t}$ is holomorphic for t in a sector independant of $\lambda \in D_o$.

iv) $\exists c > o$ and $\alpha < 1$ such that $\forall \lambda \in D_o$

$$\| e^{-L_\lambda t} \|_{\mathcal{L}(K ; \mathcal{D})} \leq c t^{-\alpha} , \quad t \in]o, T] \quad , \quad T < \infty .$$

It is known, in the case of (1) , that $\alpha = 3/4$.

v) $u \mapsto M(u)$ is analytic from $\mathcal{D}$ into K and M is real.
Moreover, we have :

$$\exists \gamma > o \; , \; \|M(u)\|_K \leq \gamma \|u\|^2_{\mathcal{D}} \qquad \forall u \in \mathcal{D} \; .$$

For more generality we can assume that this inequality holds only in a neighborhood of o in $\mathcal{D}$.

These properties ensure us the solvability of the Cauchy problem (1) with $u(o) = u_o \in \mathcal{D}$ (see [4]) in a finite length of time.

4. Now, what is our problem ? Many problems in hydrodynamics are such that one knows a steady flow v_o which is stable for $\lambda < \lambda_o$, but which becomes unstable when λ crosses a critical value λ_o . This means that for $\lambda < \lambda_o$ there is no eigenvalue of L_λ with a negative real part and that, when λ crosses λ_o, some eigenvalues of L_λ cross the imaginary axis towards the left side. In many problems, there are a great number of eigenvalues which cross, nearly at the same time, the imaginary axis, because of a special geometry of the physical problem (for example a length is much greater than other dimensions of Ω). It is classical to interpret a special kind of turbulence, which then occurs for λ near λ_o , by the fact of this great amount of eigenvalues of L_λ on the left side [15] .

Here we study a model case when 4 eigenvalues of L_λ cross the imaginary axis at the same time. The case of only two (conjugated) eigenvalues is the classical one of the Hopf bifurcation [2], [4], [8], [9] , [15], where it is known that, in general, near λ_o and only on one side there is a bifurcated periodic solution, stable only if it exists for $\lambda > \lambda_o$.

II - Periodic bifurcated solutions

1. Assumptions and notations

Let us assume

H.1 There are only 4 simple eigenvalues ($\pm i\,\omega_o$ and $\pm i\,\omega_1$) of L_{λ_o} ($\lambda_o \in \mathbb{R}$) on the imaginary axis. The remaining of the spectrum of L_{λ_o} is in the real positive side (strictly).

By the perturbation theory [11] , because of the compactness of the resolvent, we know that in a neighborhood $\mathcal{V}(\lambda_o)$ of λ_o there exists two analytic simple eigenvalues of L_λ : $\lambda \mapsto \zeta_o(\lambda)$ and $\lambda \mapsto \zeta_1(\lambda)$ such that

$$\zeta_o(\lambda) = i\,\omega_o + (\lambda - \lambda_o)\,\zeta_o^{(1)} + O(\lambda - \lambda_o)^2$$
$$\zeta_o(\lambda) = i\,\omega_1 + (\lambda - \lambda_o)\,\zeta_1^{(1)} + O(\lambda - \lambda_o)^2 ,$$

for $\lambda \in \mathcal{V}(\lambda_o)$. Now we assume in the following :

H.2 $\quad \mathrm{Re}\ \zeta_o^{(1)} < o \quad$ and $\quad \mathrm{Re}\ \zeta_1^{(1)} < o$.

This assumption ensures us that for $\lambda < \lambda_o$, all the spectrum of L_λ lies on the right side of the complex plane, which gives the Liapunov stability of the basic flow. Now, when λ crosses λ_o, 4 eigenvalues of $L_\lambda(\zeta_o, \bar{\zeta}_o, \zeta_1, \bar{\zeta}_1)$ cross at the same time the imaginary axis towards the left side. For $\lambda \in \mathcal{V}(\lambda_o)$, the remaining of the spectrum of L_λ stays in the right side.

In all the following we denote by $u^{(o)}, u^{(1)} \in \mathcal{D}$ and $w^{(o)}, w^{(1)} \in \mathcal{D}^*$ the vectors such that

$$L_{\lambda_o} u^{(o)} = i\omega_o u^{(o)} \quad , \quad L^*_{\lambda_o} w^{(o)} = -i\omega_o w^{(o)} \quad , \quad (u^{(o)}, w^{(o)})_H = 1 \; ,$$

$$L_{\lambda_o} u^{(1)} = i\omega_1 u^{(1)} \quad , \quad L^*_{\lambda_o} w^{(1)} = -i\omega_1 w^{(1)} \quad , \quad (u^{(1)}, w^{(1)})_H = 1 \; ,$$

where $L^*_{\lambda_o}$ is the adjoint of L_{λ_o} in H, of domain $\mathcal{D}^*$. Moreover we assume, without loss of generality, that $o < \omega_o < \omega_1$

2. Classical periodic solution

. 1. In a classical way, when we look for a periodic solution, we make a rescaling of t such that the period becomes 2π. If τ is the unknown period, we pose $s = 2\pi t/\tau$, then

$$u(t) = u(\tau s/_{2\pi}) = \tilde{u}(s) \; .$$

For $\tilde{u}$ we have now

$$(3) \quad \begin{cases} \dfrac{d\tilde{u}}{ds} + \eta L_\lambda \tilde{u} - \eta M(\tilde{u}) = o \; , \quad \text{where } \eta = \tau/_{2\pi} \\ \tilde{u}(s + 2\pi) = \tilde{u}(s) \end{cases}$$

where we have to specify the functional space where $\tilde{u}$ lies

Let us denote by $H^m(T ; \mathfrak{h})$ the Sobolov space of nearly everywhere 2π-periodic functions taking values in the Hilbert space $\mathfrak{h}$, such that

$$\|v\|^2_{H^m_\mathfrak{h}} = \sum_{k=o}^{m} \int_o^{2\pi} \left\| \frac{d^k v}{ds^k} \right\|^2_\mathfrak{h} ds < \infty \; .$$

The space $H^m(T ; \mathfrak{h})$ has an obvious hilbertian structure. Now, we look for solutions of (3) in $H^1(T ; \mathcal{D}) \cap H^2(T ; H)$. This space is fine because we know ([3] chap. X) that $u \mapsto M(u)$ can be extended as an analytic map from $H^1(T ; \mathcal{D})$ into $H^1(T ; K) \hookrightarrow H^1(T ; H)$.

. 2. Let us consider the linear problem

$$(4) \quad \begin{cases} \dfrac{d\tilde{u}}{ds} + \eta L_\lambda \tilde{u} = v \in H^1(T ; H) \; , \\ \tilde{u} \in H^1(T ; \mathcal{D}) \cap H^2(T ; H) \; . \end{cases}$$

We can write

$$\tilde{u}(s) = \sum_{n \in Z} u_n e^{nis} \quad , \quad v(s) = \sum_{n \in Z} v_n e^{nis} \; ,$$

where
$$\sum_{n\in Z} (1+n^2)\ \|v_n\|_H^2 < +\infty$$
and
$$\sum_{n\in Z} \left((1+n^2)\|u_n\|^2 + n^4\ \|u_n\|_H^2\right) < +\infty .$$
Hence, $\forall\, n \in Z$, we have
$$\left(\frac{ni}{\eta} + L_\lambda\right)u_n = \frac{1}{\eta}\, v_n .$$

We know, thanks to the properties I.3. ii) and iii) of L_λ, that if $-ni/\eta$ is not an eigenvalue of L_λ, we have an estimate
$$\left\|\left(L_\lambda + \frac{ni}{\eta}\right)^{-1}\right\|_{\mathcal{L}(H\,;\,\mathcal{D})} \leq c ,$$
$$\left\|\left(L_\lambda + \frac{ni}{\eta}\right)^{-1}\right\|_{\mathcal{L}(H)} \leq \frac{c}{|n|}$$

for large n, and a positive constant c. We then obtain the

Lemma 1

If $\forall\, n \in Z$, ni/η is not an eigenvalue of L_λ then O is an isolated solution of (3) in $H^1(T\,;\mathcal{D}) \cap H^2(T\,;H)$.

Proof The linearized operator has a bounded inverse from $H^1(T\,;H)$ into $H^1(T\,;\mathcal{D}) \cap H^2(T\,;H)$, hence the method of proof for this Lemma is exactly the same as for the classical theorem in the steady case. ∎

3. We know that $\pm i\,\omega_0$ and $\pm i\,\omega_0$ are the pure imaginary eigenvalues of L_{λ_0}, then we can hope to find a bifurcated cycle for λ near λ_0 only with η near n_0/ω_0 or n_1/ω_1 for any n_0, $n_1 \in \mathbb{N}^*$.

Let us take η near $\eta_0 = \omega_1^{-1}$, then the classical theory [3] runs well because $\forall\, n \in Z$ the operator $ni\,\omega_1 + L_{\lambda_0}$ is invertible except for two $n : \pm 1$. We then obtain a solution of (3):
$$s \mapsto \tilde{\mathcal{U}}_1(s,\varepsilon) = \sum_{n\geq 1} \varepsilon^n\, \mathcal{U}_1^{(n)}(s) \quad , \text{ where}$$
(5)
$$\varepsilon = |\lambda - \lambda_0|^{\frac{1}{2}} \quad , \quad \tau_1 = 2\pi\,\omega_1^{-1} + O(\varepsilon^2) ,$$
$$\mathcal{U}_1^{(1)}(s) = a_1\, e^{-is}\, u^{(1)} + \bar{a}_1\, e^{is}\, \bar{u}^{(1)} \quad , \quad |a_1|^2 = \left|\frac{\operatorname{Re}\zeta_1^{(1)}}{\gamma_1}\right|$$

only for $\lambda \geq \lambda_0$ or $\lambda \leq \lambda_0$ following $\gamma_1 < 0$ or > 0 where
$$\gamma_1 = \operatorname{Re}\Big\{ \Big(4M^{(2)}\big[u^{(1)},\ L_{\lambda_0}^{-1}\, M^{(2)}(u^{(1)}, u^{(1)})\big] + 2M^{(2)}\big[\bar{u}^{(1)}, (L_{\lambda_0} - 2i\,\omega_1)^{-1} M^{(2)}(u^{(1)}, u^{(1)})\big] + 3\, M^{(3)}\,(u^{(1)}, u^{(1)}, \bar{u}^{(1)}),\ w^{(1)}\Big)_H \Big\} ,$$

and $M(u) : \sum_{n\geq 2} M^{(n)}(u,\dots,u)$ is the Taylor series of M near O, $M^{(n)}$ is n-linear, bounded and symetric.

In the same way, we can take η near ω_o^{-1}, but to apply the classical theory, we have to assume $\omega_1 \neq p\,\omega_o \ \forall\ p \in \mathbb{N}$. Now, $n i \omega_o + L_{\lambda_o}$ is invertible except for $n = \pm 1$, hence we obtain a solution of (3).

$$s \mapsto \tilde{\mathcal{U}}_o(s,\varepsilon) = \sum_{n\geq 1} \varepsilon^n\, \mathcal{U}_o^{(n)}(s) \quad , \quad \text{where}$$

(6)

$$\varepsilon = |\lambda - \lambda|^{\frac{1}{2}}, \quad \tau_o = 2\pi\, \omega_o^{-1} + O(\varepsilon^2),$$

$$\mathcal{U}_o^{(1)}(s) = a_o\, e^{-is}\, u^{(o)} + \bar{a}_o\, e^{is}\, u^{(o)}, \quad |a_o|^2 = \left| \frac{\mathrm{Re}\,\zeta_o^{(1)}}{\gamma_o} \right|$$

only for $\lambda \geq \lambda_o$ or $\lambda \leq \lambda_o$ following $\gamma_o < o$ or $\gamma_o > o$, where γ_o is analogous to γ_1 with $u^{(o)}$, ω_o instead of $u^{(1)}$, ω_1.

This leads to the

Lemma 2

If $\omega_1 \neq p\,\omega_o$ ($\omega_1 > \omega_o$) $\forall\ p \in \mathbb{N}$, and if γ_o and $\gamma_1 \neq o$, then there exist at least two distinct periodic one-sided bifurcated solutions of (2), denoted $t \mapsto \mathcal{U}_o(t,\varepsilon)$, $t \mapsto \mathcal{U}_1(t,\varepsilon)$, where $\varepsilon = |\lambda - \lambda|^{\frac{1}{2}}$ is the order of these solutions near λ_o. We have $\mathcal{U}_k(t,\varepsilon) = \tilde{\mathcal{U}}_k\left(\frac{2\pi t}{\tau_k(\varepsilon)}, \varepsilon\right)$, $k = o, 1$, where $\tilde{\mathcal{U}}_k(s,\varepsilon)$ is defined by (5), (6).

3. Other periodic solutions

.1. Now, we can show easily the

Lemma 3

If $\omega_1/\omega_o \notin \mathbb{Q}$, and if γ_o and $\gamma_1 \neq o$, then the previous bifurcated periodic solutions of (2), $\mathcal{U}_o$ and $\mathcal{U}_1$, are the only ones which bifurcate from λ_o.

Proof Let us take η near $\eta_o = q/\omega_o$, then $\forall\ n \in \mathbb{Z}$ $L_{\lambda_o} + \frac{n i \omega_o}{q}$ is invertible, except for $n = \pm q$. The classical theory applies and we obtain a periodic bifurcated solution of period τ near $n\,\tau_o$, with the same principal part. By the uniqueness of the solution with period near $n\tau_o$. (as $\mathcal{U}_o$) we have necessarily the same solution as $\mathcal{U}_o$, with $\tau = n\,\tau_o$. The same argument is valid for η near $\eta_o = p/\omega_1$. The lemma is proved. ▮

.2. Let us look for other periodic solutions than $\mathcal{U}_1$ and $\mathcal{U}_o$ (when this one exists) in the case when $\omega_1/\omega_o = p/q \in \mathbb{Q}$, $p > q$. We take here η near $\eta_o = \frac{q}{\omega_o} = \frac{p}{\omega_1}$. This η_o gives the new fact that the operator $L_{\lambda_o} + \frac{n i \omega_o}{q}$ is not invertible for $n = \pm q$ and $n = \pm p$. All the possible η_o which give this situation are multiple of this one, and will give the same solutions, but with an artificial multiple period.

Let us now define $\mathbb{H} = H^1(T;H)$, $\mathbb{D} = H^1(T;\mathcal{D}) \cap H^2(T;H)$ and the projection $\mathbf{P}_o$: $\forall\, u \in L^2(T;H)$,

$$(\mathbf{P}_o u)(s) = \alpha_o\, e^{-qis}\, u^{(o)} + \beta_o\, e^{qis}\, \bar{u}^{(o)} + \alpha_1\, e^{-pis}\, u^{(1)} + \beta_1\, e^{pis}\, \bar{u}^{(1)} ,$$

where $\alpha_o = \frac{1}{2\pi}\int_o^{2\pi} e^{qis}\left(u(s), w^{(o)}\right)_H ds$, $\beta_o = \frac{1}{2\pi}\int_o^{2\pi} e^{-qis}\left(u(s), \bar{w}^{(o)}\right)_H ds$,

$\alpha_1 = \frac{1}{2\pi}\int_o^{2\pi} e^{pis}\left(u(s), w^{(1)}\right)_H ds$, $\beta_1 = \frac{1}{2\pi}\int_o^{2\pi} e^{-pis}\left(u(s), \bar{w}^{(1)}\right)_H ds$.

We have $\mathbf{P}_o \in \mathcal{L}\left(L^2(T;H)\ ;\ \mathbb{D}\right)$ and $\mathbf{P}_o^2 = \mathbf{P}_o$, and the

<u>Lemma 4</u>

Let us define the operator T_o , of domain $\mathbb{D}$, in $\mathbb{H}$, such that $\forall\, u \in \mathbb{D}$,

$$(T_o u)(s) = \frac{du}{ds}(s) + \eta_o L_{\lambda_o} u(s) .$$

Then T_o is a closed operator in $\mathbb{H}$, and admits an adjoint T_o^* such that $\mathbb{D}^* = H^1(T;\mathcal{D}^*) \cap H^2(T;H)$ is the domain of T_o^* and $\forall\, v \in \mathbb{D}$,

$$(T_o^* v)(s) = -\frac{dv}{ds}(s) + \eta_o L_{\lambda_o}^* v(s) .$$

Moreover 0 is a semi-simple eigenvalue of multiplicity 4 , the kernel of T_o being the range of $\mathbf{P}_o$ and we have $\mathbf{P}_o T_o \subset T_o \mathbf{P}_o$.

<u>Proof</u> The adjoint is easy to find in a classical way, once it is proved that T_o is a closed densely defined operator. $\mathbb{D}$ is dense in $\mathbb{H}$ because of the density of $H^2(T;\mathcal{D})$ in $\mathbb{H}$ thanks to the density of $\mathcal{D}$ in H . The operator T_o is closed because it can be shown that the problem

$$(7)\quad \begin{cases} \dfrac{du}{ds} + \eta_o L_{\lambda_o} u = v \in \mathbb{H} , \\ u \in \mathbb{D} , \end{cases}$$

has a solution if and only if $\mathbf{P}_o v = o$. In this case, there is a unique solution such that $\mathbf{P}_o u = o$. We denote by $R_o : (\mathbb{1} - \mathbf{P}_o)\mathbb{H} \to (\mathbb{1} - \mathbf{P}_o)\mathbb{D}$ the linear operator such that $u = R_o v$ satisfies $\mathbf{P}_o u = o$ solution of (7). Hence $R_o \in \mathcal{L}[(\mathbb{1}-\mathbf{P}_o)\mathbb{H}\ ;\ (\mathbb{1}-\mathbf{P}_o)\mathbb{D}]$. The relation $\mathbf{P}_o T_o \subset T_o \mathbf{P}_o$ is easy to verify. ∎

3. Let us consider the system (3) with η near $\eta_o = q/\omega_o = p/\omega_1$, where $\tilde{u} \in \mathbb{D}$. We rewrite it into the form :

$$(8)\quad \begin{cases} T_o \tilde{u} = N(\lambda, \eta, \tilde{u}) \\ \tilde{u} \in \mathbb{D} , \end{cases}$$

where $N(\lambda, \eta, \tilde{u}) = (\eta_o L_{\lambda_o} - \eta L_\lambda)\tilde{u} + \eta M(\tilde{u}) \in \mathbb{H}$. Let us pose $\tilde{u} = x + y$

with $x = P_\circ \tilde{u}$, $y = (1 - P_\circ)\tilde{u}$. The decomposition of (8) gives

(9) $$P_\circ N(\lambda, \eta, x + y) = 0 ,$$

(10) $$y = R_\circ(1 - P_\circ) N(\lambda, \eta, x + y) .$$

In a classical way, the equation (10) can be solved by the implicit function theorem, with respect to y . We find an analytic y :

$$y = \mathfrak{R}(\lambda, \eta, x) = \sum_{\substack{r \geq 1 \\ m+n+r \geq 2}} (\lambda - \lambda_\circ)^m (\eta - \eta_\circ)^n \mathfrak{R}^{(r)}_{mn} [x^{(r)}],$$

with $\mathfrak{R}^{(1)}_{01} = 0$, when $\mathfrak{R}^{(r)}_{mn}$ is bounded r-linear and symmetric. The series converges for $|\lambda - \lambda_\circ|$, $|\eta - \eta_\circ|$, $\|x\|$ small enough. Now, we write

$$x(s) = \alpha_\circ e^{-qis} u^{(0)} + \bar{\alpha}_\circ e^{qis} \bar{u}^{(0)} + \alpha_1 e^{-pis} u^{(s)} + \bar{\alpha}_1 e^{pis} \bar{u}^{(1)} ,$$

then it can be shown that

$$(\mathfrak{R}^{(r)}_{mn} [x^{(r)}])(s) = \sum_{k_1+k_2+k_3+k_4=r} \alpha_\circ^{k_1} \bar{\alpha}_\circ^{k_2} \alpha_1^{k_3} \bar{\alpha}_1^{k_4} e^{[(k_2 - k_1)q + (k_4 - k_3)p]is} v_{k_1 k_2 k_3 k_4},$$

where $v_{k_1 k_2 k_3 k_4} \in \mathcal{D}$. The bifurcation system is obtained by replacing y in (9) by $\mathfrak{R}(\lambda, \eta, x)$. Hence we have a system of two complex equations :

(11) $$\int_0^{2\pi} e^{qis} \left(N[\lambda, \eta, x(s) + \{\mathfrak{R}(\lambda, \eta, x)\}(s)], w^{(0)}\right)_H ds = 0$$

(12) $$\int_0^{2\pi} e^{pis} \left(N[\lambda, \eta, x(s) + \{\mathfrak{R}(\lambda, \eta, x)\}(s)], w^{(1)}\right)_H ds = 0 .$$

The unknown are $\alpha_\circ$, α_1, η which are functions of λ , i.e. 5 unknown for 4 equations. The undetermination on the origin of s will allow us to solve the system . The equation (11) gives

(13) $$\sum (\lambda - \lambda_\circ)^m (\eta - \eta_\circ)^n \alpha_\circ^{k_1} \bar{\alpha}_\circ^{k_2} \alpha_1^{k_3} \bar{\alpha}_1^{k_4} \beta^{(m,n)}_{k_1 k_2 k_3 k_4} = 0 ,$$

where the summation is on $k_i \geq 0$, $k_1 + k_2 + k_3 + k_4 = r \geq 1$, $m + n + r \geq 2$, and $(k_2 - k_1 + 1)q + (k_4 - k_3)p = 0$.

The equation (12) gives

(14) $$\sum (\lambda - \lambda_\circ)^m (\eta - \eta_\circ)^n \alpha_\circ^{k_1} \bar{\alpha}_\circ^{k_2} \alpha_1^{k_3} \bar{\alpha}_1^{k_4} \gamma^{(m,n)}_{k_1 k_2 k_3 k_4} = 0 ,$$ where the

summation is on $k_i \geq 0$, $k_1 + k_2 + k_3 + k_4 = r \geq 1$, $m+n+r \geq 2$, and $(k_2 - k_1)q + (k_4 - k_3 + 1)p = 0$.

Now, it is easy to see that $|\alpha_\circ|$ is in factor in (13), whereas $|\alpha_1|$ is in factor in (14) if $q \neq 1$. Now we have

$$\beta^{(0,1)}_{1000} = -i\omega_\circ , \quad \beta^{(1,0)}_{1000} = -\eta_\circ \zeta_\circ , \quad \beta^{(0,0)}_{k_1, k_2, k_3, k_4} = 0 \text{ for } k_1 + k_2 + k_3 + k_4 = 2$$

except $(k_1, k_2, k_3, k_4) = (o, 1, 1, o)$ when $q = 1$, $p = 2$. In the same way

$$\gamma^{(\omega,1)}_{oo1o} = -i\omega_1, \quad \gamma^{(o,1)}_{oo1o} = -\eta_\circ \zeta_1^{(1)}, \quad \gamma^{(o,o)}_{k_1 k_2 k_3 k_4} = o \quad \text{for} \quad k_1+k_2+k_3+k_4 = 2$$

except $(k_1, k_2, k_3, k_4) = (2, o, o, o)$ when $q = 1$, $p = 2$. For $\omega_1 = 2\omega_\circ$, we have

$$\beta^{(o,o)}_{o11o} = 2\eta_\circ (M^{(2)}(\overline{u}^{(o)}, u^{(1)}), w^{(o)})_H ,$$

$$\gamma^{(o,o)}_{2ooo} = \eta_\circ (M^{(2)}(u^{(o)}, u^{(o)}), w^{(1)})_H .$$

4. The solution $\alpha_\circ = o$

The equation (13) is satisfied and it remains (14) of the form $g(\alpha_1, \eta, \lambda) = o$ in $\mathbb{C}$, where the argument of α_1 is undetermined. It is the classical case. This leads to the following alternative :

either $\alpha_1 = o$, then $x = o$ and $\tilde{u} = o$ (trivial solution)

or $\alpha_1 \neq o$, then $\tilde{u}$ is exactly $\tilde{\mathcal{U}}_1$ after the change $qs \to s$. Hence we have obtained the bifurcated cycle $\mathcal{U}_1$ plus the trivial solution.

5. Case $\omega_1/\omega_\circ = p/q$ with $q \neq 1$.

Now, we have to consider the solution $\alpha_1 = o$ which solves (14). The same argument as previously shows that we obtain, either the trivial solution or the bifurcated cycle $\mathcal{U}_\circ$. Let us now assume $\alpha_\circ > o$ in the aim to eliminate the undetermination on the origin of s. We have then (13) (14) into the form :

(15) $$\alpha_\circ f(\alpha_\circ, \alpha_1, \eta, \lambda) = o ,$$

(16) $$g(\alpha_\circ, \alpha_1, \eta, \lambda) = o ,$$

where $|\alpha_1|$ is in factor in (16). The equation (15) in $\mathbb{C}$ can be solved with respect to (λ, η) by the implicit function theorem :

(17) $$\begin{cases} \lambda - \lambda_\circ = \Lambda(\alpha_\circ, \alpha_1) = \lambda_{20}\alpha_\circ^2 + \lambda_{21}|\alpha_1|^2 + O(|\alpha_\circ| + |\alpha_1|)^3 , \\ \eta - \eta_\circ = \Theta(\alpha_\circ, \alpha_1) = \theta_{20}\alpha_\circ^2 + \theta_{21}|\alpha_1|^2 + O(|\alpha_\circ| + |\alpha_1|)^3 , \end{cases}$$

where Λ and Θ are analytic. Replacing in (16), we obtain an equation of the form $G(\alpha_\circ, \alpha_1) = o$:

(18) $$\alpha_1 \Big[\Big(\gamma^{(o,o)}_{oo21} + \lambda_{21}\gamma^{(1,o)}_{oo1o} + \theta_{21}\gamma^{(o,1)}_{oo1o}\Big)|\alpha_1|^2 +$$

$$+ \left(\gamma_{1110}^{(0,0)} + \gamma_{0010}^{(1,0)} \lambda_{20} + \gamma_{0010}^{(0,1)} \theta_{20}\right)\alpha_0^2\Big] +$$

$$+ O\left[|\alpha_1| \left(|\alpha_0| + |\alpha_1|\right)^3\right] = 0 .$$

Eliminating the solution $\alpha_1 = 0$ which was studied previously, we obtain, for the principal part of (α_0, α_1) an equation of the type $A|\alpha_1|^2 + B\,\alpha_0^2 = 0$, where A and $B \in \mathbb{C}$. In general the only solution is the O solution. This proves the

Theorem 1

If $\omega_1 \neq p\,\omega_0$ $\forall\ p \in \mathbb{N}$ $(\omega_1 > \omega_0)$ and if γ_0 and $\gamma_1 \neq 0$, then in general, the two distinct, periodic, one-sided bifurcated solutions of (2), $\mathcal{U}_0$ and $\mathcal{U}_1$, are the only ones which bifurcate from λ_0.

This is more precise statement than those of the Lemmas 2 and 3 .

6. Case $\omega_1 = p\ \omega_0$ $(p \geq 2)$

The system (15) (16) holds, and the solution $\alpha_0 = 0$ corresponds to the bifurcated cycle $\mathcal{U}_1$. Moreover (17) holds if $p \geq 3$, and the principal part of (18) holds if $p \geq 4$, the remaining part being $O(|\alpha_0| + |\alpha_1|)^4$.

Let us assume $p \geq 4$, then the lower order term purely in α_0 is $\gamma_{p000}^{(0,0)}\ \alpha_0^p$. The study of the Newton diagram and the previous study of the principal part of the form $A\ \alpha_1\ |\alpha_1|^2 + B\alpha_1\ \alpha_0^2$ gives us the only other solution to be considered : $\alpha_1 = \alpha_0^{(p-2)}\ z(\alpha_0)$ where $z(0)$ is in general uniquely determined by a linear equation. It is then classical to find z as an analytic function of α_0, which gives α_1 analytic in α_0 and the

Theorem 2

If $\omega_1 = p\,\omega_0$, $p \geq 4$, $p \in \mathbb{N}$, and if γ_0 and $\gamma_1 \neq 0$, then there exist two distinct periodic one-sided bifurcated solutions of (2), $\mathcal{U}_1$ and $\mathcal{U}_0'$. The solution $\mathcal{U}_1$ is the classical previous one, whereas the other, $\mathcal{U}_0'$ is analogous to $\mathcal{U}_0$ (which does not exist in general here) in the sense that it is of order $|\lambda - \lambda_0|^{\frac{1}{2}}$ and has the same principal part as (6) .

Let us assume now that $\omega_1 = 2\,\omega_0$. Then, we have now, instead of (17)

$$(17')\qquad \begin{cases} \Lambda(\alpha_0, \alpha_1) = \Lambda_1\ \alpha_1 + \overline{\Lambda}_1\ \overline{\alpha}_1 + O(|\alpha_0| + |\alpha_1|)^2 , \\ \Theta(\alpha_0, \alpha_1) = \Theta_1\ \alpha_1 + \overline{\Theta}_1\ \overline{\alpha}_1 + O(|\alpha_0| + |\alpha_1|)^2 . \end{cases}$$

The principal part of $G(\alpha_o, \alpha_1)$ is of order 2, in general :

$$\alpha_1 \left[\gamma_{oo1o}^{(1,o)} (\Lambda_1 \alpha_1 + \bar{\Lambda}_1 \bar{\alpha}_1) + \gamma_{oo1o}^{(o,1)} (\Theta_1 \alpha_1 + \bar{\Theta}_1 \bar{\alpha}_1)\right] + \gamma_{2ooo}^{(o,o)} \alpha_o^2 .$$

Let us pose $\alpha_1 = \alpha_o z(\alpha_o)$, and eliminate the solution $\alpha_o = o$, we obtain for $z(o)$ an elementary problem which has either no solution, or 4 solutions : $z_1(o)$, $-z_1(o)$, $z_2(o)$, $-z_2(o)$ (we do not consider the double solutions, where the implicit function theorem cannot be applied). In fact, two opposite solutions $z_i(o)$ give the same cycle because of a translation on s . Then we have the

Theorem 3

If $\omega_1 = 2\omega_o$, and if $\gamma_1 \neq o$, then we have the one-sided bifurcated solution $\mathcal{U}_1$ of order $|\lambda - \lambda_o|^{\frac{1}{2}}$. Moreover, following the coefficients, either it is the only bifurcating solution, or there exist two other non trivial, two-sided, bifurcating solutions $\mathcal{U}_2$ and $\mathcal{U}_2'$ of order $(\lambda - \lambda_o)$, whose period is near $2\tau_1$.

It remains now to study the case when $\omega_1 = 3\omega_o$. The relations (17) hold but (18) does not hold because of the additional term $\gamma_{3ooo}^{(o,o)} \alpha_o^3$. Let us pose $\alpha_1 = \alpha_o z(\alpha_o)$ and eliminate the solution $\alpha_o = o$. For $z(o)$ we obtain an equation of the type

$$A\, z\, |z|^2 + B\, z + \gamma_{3ooo}^{(o,o)} = o ,$$

with complex coefficients, which admits following the coefficients 1, 3, 5, or 7 simple solutions. In the same way as previously, we obtain the

Theorem 4

If $\omega_1 = 3\omega_o$, and if $\gamma_1 \neq o$, then we have the one-sided bifurcated solution $\mathcal{U}_1$ of order $|\lambda - \lambda_o|^{\frac{1}{2}}$. Moreover, following the coefficients, there exist 1 or 3 or 5 or 7 distinct, non trivial, one-sided other bifurcated solutions, all of order $|\lambda - \lambda_o|^{\frac{1}{2}}$, whose period is near $3\tau_1$.

III - Invariant torus and stability of the solutions

1. Reduction to a 4-dimensional problem

Let us assume

[H.3 $\omega_1 \neq p\,\omega_o$ for $p = 2$ or 3 $(\omega_1 > \omega_o > o)$.

We know by theorems 1 and 2 that there exist two one-sided bifurcating periodic solutions denoted $\mathcal{U}_1$, $\mathcal{U}_o$ (or $\mathcal{U}_o'$) for λ near λ_o. Our aim is now to look for an invariant manifold of dimension 2 for the dynamical system (2).

We know (see [3] or [4]) that the map

$$(\lambda, u_o) \mapsto \Psi_\lambda(u_o) = \mathcal{U}(T, \lambda, u_o)$$

is analytic from $\mathbb{C} \times \mathcal{D}$ into $\mathcal{D}$ for $T \in]o, \infty[$, $\lambda \in \mathcal{O}$ open $\Subset D_o$ (included in a compact of D_o), $u_o \in \mathcal{V}(o) \subset \mathcal{D}$, where $t \mapsto \mathcal{U}(t, \lambda, u_o)$ is the solution, continuous on $[o, T_o]$ into $\mathcal{D}$, of the Cauchy problem (2), with $u(o) = u_o \in \mathcal{D}$.

The derivative of Ψ_λ at O is

$$D\Psi_\lambda(O) = e^{-L_\lambda T}$$

and it is known that this operator is compact because of the property ii) of L_λ (see [4] or [14]). Now, by the assumption H.1 and H.2 we know that for $\lambda \in \mathcal{V}^-(\lambda_o)$ (left real neighborhood of λ_o) all eigenvalues of L_λ are of positive real part, which gives the stability of the basic flow, whereas for $\lambda \in \mathcal{V}^+(\lambda_o)$, 4 eigenvalues of L_λ are of negative real part, the remaining of the spectrum being of positive real part.

It is easy to see that any eigenvalue $\zeta \neq o$ of $e^{-L_\lambda T}$ corresponds to a determination of $\frac{1}{T} \log \zeta$ as an eigenvalue of $-L_\lambda$, the eigenvectors being the same.

Moreover, we know by [3], that $(e^{-L_\lambda T})^* = e^{-L_\lambda^* T}$, hence we here have the eigenvectors $u^{(o)}, \bar{u}^{(o)}, u^{(1)}, \bar{u}^{(1)}$ respectively for the eigenvalues $e^{-i\omega_o T}, e^{i\omega_o T}, e^{-i\omega_1 T}, e^{i\omega_1 T}$ of $e^{-L_{\lambda_o} T}$.

Moreover we have the eigenvector $w^{(o)}, \bar{w}^{(o)}, w^{(1)}, \bar{w}^{(1)}$ respectively for the eigenvalues $e^{i\omega_o T}, e^{-i\omega_o T}, e^{i\omega_1 T}, e^{-i\omega_1 T}$ of $e^{-L^*_{\lambda_o} T}$.

This leads to the fact that, if they are distinct, all these eigenvalues are simple of modulus 1, the remaining of the spectrum of $e^{-L_{\lambda_o} T}$ being strictly inside the unit disc.

Let us now consider the map : $\mathcal{D} \times \mathbb{C} \to \mathcal{D} \times \mathbb{C}$

$$\Phi : [u_o, \lambda] \mapsto [\Psi_\lambda(u_o), \lambda]$$

which is analytic for $u_o \in \mathcal{V}(O) \subset \mathcal{D}$, $\lambda \in \mathcal{O} \Subset D_o$, $T \in]o, \infty[$.

The derivative at the point (O, λ_o) is such that $\forall [v, \mu] \in \mathcal{D} \times \mathbb{C}$

$$D\Phi(O, \lambda_o)[v, \mu] = [e^{-L_{\lambda_o} T} v, \mu],$$

and its spectrum consists with a part on the unit circle : the simple eigenvalues $e^{\pm i\omega_o T}$, $e^{\pm i\omega_1 T}$ and 1. The remaining part of the spectrum is strictly inside the unit disc. Let us note E the projection relative to the eigenvalues $e^{\pm i\omega_o T}$, $e^{\pm i\omega_1 T}$ of $e^{-L_{\lambda_o} T}$, commuting with this operator, and let us apply the Center-manifold theorem (see [13] or [7]) :

there exist a neighborhood $\mathcal{O}$ of (O, λ_o) in $\mathcal{D} \times \mathbb{R}$, and a regular 5-dimensional sub-manifold $M \subset \mathcal{O}$, passing through (O, λ_o), tangent to $E\mathcal{D} \times \mathbb{R}$ at this point, such that

i) M is "locally invariant" by Φ,

ii) M is "locally attracting".

The section $M_{\lambda,T}$ of M in $\mathcal{B}$ is a 4-dimensional manifold, its equation being

(19) $$y = G_T(x, \lambda), \quad x \in E\,\mathcal{B}, \quad y \in (\mathbb{1} - E)\,\mathcal{B} .$$

Moreover it can be shown that :

$$G_T(O, \lambda_\circ) = O, \quad \frac{\partial G_T}{\partial x}(O, \lambda_\circ) = O, \quad \frac{\partial G_T}{\partial \lambda}(O, \lambda_\circ) = O \text{ and}$$

$$G_T(O, \lambda) = O \quad \text{for } \lambda \in \mathcal{V}(\lambda_\circ) \subset \mathbb{R} \quad \text{because of} \quad \mathcal{U}(T, \lambda, O) = O .$$

In fact, we can show the important

Lemma 5

The locally invariant 4-dimensional manifold $M_{\lambda,T}$ is independant of T, and locally invariant by the dynamical system (2) .

Proof We can choose an open set $\mathcal{O}_1$ in $\mathcal{B}$ such that

i) $\forall u_\circ \in \mathcal{O}_1$ then $\mathcal{U}(T, \lambda, u_\circ) \in \mathcal{O}$ where $\mathcal{O}$ is such that if $u_\circ \in M_{\lambda,T}$ then $\mathcal{U}(T, \lambda, u_\circ) \in M_{\lambda,T}$.

ii) $\forall t_\circ \in [o, T]$, if $u_1 \in M_{\lambda,T,t_\circ} = \{u \in \mathcal{O}_1 ;\ \exists\, u_\circ \in M_{\lambda,T}$ such that $\mathcal{U}(t_\circ, \lambda, u_\circ) = u\}$, then $\exists\, u_\circ$ unique in $M_{\lambda,T}$ such that $u_1 = \mathcal{U}(t_\circ, \lambda, u_\circ)$.

The existence of $\mathcal{O}_1$ is due to the fact that if we consider the system $u_1 = x_1 + y_1$, $x_1 = E u_1$, $u_\circ = x_\circ + y_\circ$, $x_\circ = E u_\circ$

(20) $$x_1 = E\ e^{-L_\lambda t_\circ}(x_\circ + y_\circ) + O(\|u_\circ\|^2)$$

(21) $$y_1 = (\mathbb{1} - E)\ e^{-L_\lambda t_\circ}(x_\circ + y_\circ) + O(\|u_\circ\|^2)$$

where $y_\circ = G_T(x_\circ, \lambda) = O(|\lambda - \lambda_\circ|\ \|x_\circ\| + \|x_\circ\|^2)$, then (20) can be solved with respect to $x_\circ$ as a function of x_1 and (21) gives $y_1 = G\,(x_1, \lambda, T, t_\circ) = O(|\lambda - \lambda_\circ|\,\|x_1\| + \|x_1\|^2)$. Then $M_{\lambda,T,t_\circ}$ is a 4-dimensional manifold tangent to $E\,\mathcal{B}$ at O for $\lambda = \lambda_\circ$. Moreover it is easy to show that if $u_1 \in M_{\lambda,T,t_\circ}$ and if $\mathcal{U}(T, \lambda, u_1) \in \mathcal{O}_1$ then $\exists\, u_\circ \in M_{\lambda,T}$ and $\mathcal{U}(T, \lambda, u_\circ) \in M_{\lambda,T}$ and $\mathcal{U}[t_\circ, \lambda, \mathcal{U}(T, \lambda, u_\circ)] = \mathcal{U}(T, \lambda, u_1) \in M_{\lambda,T,t_\circ}$. By the uniqueness, we then have $M_{\lambda,T,t_\circ} = M_{\lambda,T} \cap \mathcal{O}_1$, $\forall\, t_\circ \in [o, T]$. Hence, if $u_\circ \in M_{\lambda,T} \cap O_1$ then $\mathcal{U}(t, \lambda, u_\circ) \in M_{\lambda,T}$ $\forall\ t \in [o, T]$. ∎

We shall write now $G(x, \lambda)$ instead of $G_T(x, \lambda)$ because T does not occur in G . Let us define the regular map : $\mathbb{R} \times E\,\mathcal{B} \to E\,\mathcal{B}$

(22) $$(\lambda, x) \mapsto \psi(\lambda, x) = E\ \Psi_\lambda\,[x + G\,(x, \lambda)]$$

in a neighborhood of $(\lambda_\circ, O)$. Our aim is to look for an invariant manifold for the map $x \mapsto \psi(\lambda, x)$ in the 4-dimensional space $E\,\mathcal{B}$.

2. Application of the results of R. JOST and E. ZEHNDER

Let us note $\tau_0(\lambda)$ the period of the bifurcated solution $\mathcal{U}_0$(or $\mathcal{U}_0'$) and $\tau_1(\lambda)$ the period of $\mathcal{U}_1$, and let us choose $T=\tau_0(\lambda)$ in $\psi(\lambda, .)$.

Then, we already know that this map has an invariant cycle $E\,\mathcal{U}_0$ of fixed points . We have the same phenomenon when $T = \tau_1(\lambda)$ for the invariant cycle $E\,\mathcal{U}_1$. Now, for thechnical reasons we have to choose T with the minimum of rational relations with $\tau_0(\lambda)$ and $\tau_1(\lambda)$.

In fact, we have a regular map ψ such that

$$\frac{\partial\psi}{\partial x}(\lambda_0, 0) = E\ e^{-L_{\lambda_0} T} = e^{-L_{\lambda_0} T}\ E$$

in $E\,\mathcal{B}$, and we can use the work of R. JOST and E. ZEHNDER [10] , in doing in their work $\mu_1 = \mu_2 = \lambda - \lambda_0 = \mu$. Their assumptions in the theorem 1 of [10] will be fulfilled if the following property holds :

$$\begin{cases} (s_1\,\omega_0 + s_2\,\omega_1)T = 2\pi m, & s_i \in Z, \quad m \in Z, \\ |s_1| + |s_2| \leq 4 \end{cases}$$

leads to $s_1 = s_2 = m = 0$. Thanks to the assumption H.3 this is realised if we choose T such that $2\pi/T$ is not a rational combination of ω_0 and ω_1 . Hence there exists a regular changing of variables in $E\,\mathcal{B}$, leading to a normal form of the map $\psi(\lambda, \cdot)$ in $\mathbb{C}^2$. Noting $z_k = r_k\ e^{i\varphi_k}$, $Z_k = R_k\ e^{i\Phi_k}$, $k = 1, 2$, the map becomes :

$$(23)\qquad \begin{cases} R_k = (1 + \mu \pm r_k^2 + a_k(\mu)\, r_{k'}^2)\, r_k + \theta_4^{(k)}, & k' \neq k \\ \Phi_k = \varphi_k + \alpha_k(\mu) + \sum_{j=1}^{2} d_{kj}(\mu)\, r_j^2 + \theta_3^{(k)}, \end{cases}$$

here $\mu = \lambda - \lambda_0$, a_k, $\tilde{\alpha}_k$, d_{kj} are continuous functions in a neighborhood of o , and the $\theta_n^{(k)}$ are of order $(|r_1| + |r_2|)^n$, and $\tilde{\alpha}_1(0) = \omega_0 T$, $\tilde{\alpha}_2(0) = \omega_1 T$.

Let us assume

H.4 γ_0 and γ_1 , defined in (5) and (6) , are < 0 .

In this case we have

$$(23')\qquad R_k = (1 + \mu - r_k^2 + a_k(\mu)\, r_{k'}^2)\, r_k + \theta_4^{(k)}, \qquad k = 1, 2$$

and it is easy to obtain the two, previously obtained, bifurcated cycles $\mathcal{U}_0$(or $\mathcal{U}_0'$) and $\mathcal{U}_1$ for $\mu \geq 0$ (see theorems 1 and 2).

Truncating (23') by suppressing the $\theta_4^{(k)}$, we obtain the principal part of the two cycles :

$$\mathcal{U}_0 : r_1 = \mu^{\frac{1}{2}},\ r_2 = 0,$$
$$\mathcal{U}_1 : r_1 = 0,\ r_2 = \mu^{\frac{1}{2}}.$$

Stability of the cycles

It is known by [10] that the cycle $\mathcal{U}_o$ is stable (resp-unstable) if $a_2(o) < -1$ (resp. $a_2(o) > -1$; the cycle $\mathcal{U}_1$ is stable (resp-unstable) if $a_1(o) < -1$ (resp. $a_1(o) > -1$).

Now, the same truncating of (23') gives us in good conditions the principal part of an invariant two-dimensional torus :

i) if $a_1(o) < -1$ and $a_2(o) < -1$ or if $a_1(o) \cdot a_2(o) < 1$ with $a_1(o)$ and $a_2(o) > -1$ we obtain for $\mu \geq o$ ($\lambda \geq \lambda_o$)

$$\begin{cases} r_1 = \beta_1 \mu^{\frac{1}{2}} \quad , \quad r_2 = \beta_2 \mu^{\frac{1}{2}} \quad \text{with} \\ \beta_1^2 - a_1(o)\, \beta_2^2 = 1 \\ \beta_2^2 - a_2(o)\, \beta_1^2 = 1 \quad . \end{cases}$$

ii) if $a_1(o) \,.\, a_2(o) > 1$ with $a_1(o)$ and $a_2(o) > o$ we have $\mu \leq o$

$$\begin{cases} r_1 = \beta_1 (-\mu)^{\frac{1}{2}} \quad , \quad r_2 = \beta_2 (-\mu)^{\frac{1}{2}} \quad \text{with} \\ \beta_1^2 - a_1(o)\, \beta_2^2 = -1 \\ \beta_2^2 - a_2(o)\, \beta_1^2 = -1 \quad . \end{cases}$$

The complete map (23) - (23') givesitself perturbated cycles and torus when the principal part exists (see [10]).

The invariant torus for the map $\phi(\lambda, \,.\,)$ gives an invariant torus for the map Ψ_λ in $\mathcal{S}$ and in fact an invariant torus for the dynamical system (2) . Using [10] we have

Stability of the torus

when it exists, the torus is stable (resp. unstable) if $a_1(o)\; a_2(o) < 1$ (resp. $a_1(o) \,.\, a_2(o) > 1$) .

For instance, for good coefficients we can obtain for $\lambda > \lambda_o$ two unstable bifurcated cycles $\mathcal{U}_o$, $\mathcal{U}_1$ and a stable torus in $\mathcal{S}$.

$$(a_1(o) > -1 \;,\quad a_2(o) > -1 \;,\quad a_1(o) \,.\, a_2(o) < 1) \quad .$$

BIBLIOGRAPHY

[1] G.DURAND , Thèse de 3ème cycle, Pub.Math . Orsay n°128 (1975)

[2] E.HOPF , Berichten der Math -Phys.Kl.Sächs.Akad.Wiss. Leipzig 94 , 1-22 (1942)

[3] G.IOOSS , Bifurcation et Stabilité, Cours de 3ème cycle 1972-1974 , Pub.Math Orsay n°31 (1974)

[4] G.IOOSS , Arch. Rational Mech.Anal. 47 , 301-329 (1972)

[5] G.IOOSS , Communication at the IUTAM-IMU Symposium on applications of methods of functional analysis to problems of mechanics. To appear in the notes of the Symposium (Springer)

[6] G.IOOSS , Sur la bifurcation secondaire d'une solution stationnaire de systèmes du type Navier-Stokes. (En préparation)

[7] G.IOOSS , Variétés invariantes et systèmes dynamiques. To appear in the notes of the Seminar. Nice 1975

[8] V.I.IUDOVICH , Prikl.Mat.Mek. 35 , 638-655 (1971), and Prikl.Mat.Mek. 36 , 450-459 (1972)

[9] D.D.JOSEPH and D.H.SATTINGER , Arch. Rational Mech.Anal 45 , 79-109 (1972)

[10] R.JOST and E.ZEHNDER , Helvetica Physica Acta , 45 , 258 , 276 (1972)

[11] T.KATO , Perturbation theory for linear operators. Berlin-Heidelberg-New-York, Springer, 1966

[12] O.A.LADYZHENSKAYA , The mathematical theory of viscous incompressible flow. New-York , Gordon and Breach , 1963

[13] O.E.LANFORD III , Bifurcation of Periodic Solutions into Invariant Tori ... Lecture Notes in Maths , n°322 , 159-192. Berlin-Heidelberg-New-York , Springer (1973)

[14] A.PAZY , J.Math.Mech ., 17 , 12, 1131-1141 (1968)

[15] D.RUELLE and F.TAKENS , Comm.Math.Phys. 20 , 167-192 (1971)

[16] R.TEMAM , On the theory and numerical analysis of Navier-Stokes equations. Lecture Notes n°9 , University of Maryland (1973).

FACTORIZATION THEOREMS FOR THE STABILITY OF BIFURCATING SOLUTIONS

by

Daniel D. Joseph

The theory of bifurcation at a simple complex eigenvalue, developed for ordinary differential equations by Hopf (1942) and extended to partial differential equations, like the Navier-Stokes equations, by Joseph and Sattinger (1972)*, using Hopf's methods, and by Iooss (1972), Yudovich (1971), and Marsden (1973), using other methods, is a local theory which is restricted to small values of ε, the amplitude of the bifurcating solution. In the local theory, bifurcating solutions which branch to the right (supercritical bifurcation) are stable and bifurcating solutions which branch to the left (subcritical bifurcation) are unstable. I am going to derive the form which this stability result must take when the restriction on the size of the amplitude of the bifurcating solution is removed.† Subject to conditions, we are going to replace Hopf's local statement of stability with a global statement of stability. The local statement, due to Hopf, is roughly: "Subcritical solutions branching at a simple eigenvalue are unstable; supercritical solutions are stable." The global statement is: "Solutions for which the response decreases with increasing amplitude are unstable; solutions for which the response increases with amplitude are stable." Expressed in physical terms, the global statement asserts that pipe flows for which the mass flux increases as the pressure gradient decreases are unstable or, for another example, convection for which the heat transported decreases as the temperature is increased is unstable.

The results to be given here trace the eigenvalues of the Frechet derivative of the nonlinear operator whose null space contains the bifurcating solution. The main result is a factorization theorem which shows among other things that the relevant eigenvalue vanishes at critical points of the bifurcation curve. When carried to small amplitudes we recover and extend Hopf's original stability results. We do not consider secondary bifurcations here; secondary bifurcations certainly alter the stability interpretation of the theorems but not the theorems.

The recovery of stability on subcritical branches which turn around is a physically important result which may have applications to observations of the mechanics of subcritical turbulence. I will discuss

* This paper is designated in the sequel by the letters JS.

† Mathematically, the result takes form in the factorization theorems of Joseph (see Joseph & Nield, 1975). I wish to thank Professor Nield for his important contributions to the computations which at an early stage of the investigation led me to the factorization. The good suggestions of Professors P. Rabinowitz and M. Crandall about the local interpretation of the factorization are also most gratefully acknowledged.

these applications at the conclusion of this lecture.

We are now ready to state and prove our main result. Consider the following evolution problem on a Banach space:

$$\frac{dV}{dt} + L(\mu)V + N(\mu;V) = 0 \tag{1}$$

where μ is a real parameter,

$$L(\mu) = L_0 + \mu L_1 + \mu^2 L_2 + \dots$$

is a linear operator, analytic in μ, and $N(\mu;V)$ is a nonlinear operator, analytic in V and μ, whose power series in V starts with terms of at least second degree. To simplify the computations, we take

$$L(\mu) = L_0 + \mu L_1 \tag{2}$$

and consider quadratic nonlinearities

$$N(\mu,\cdot) = N(\cdot,\cdot). \tag{3}$$

Without loss of generality we shall follow JS and assume that $V(t) \equiv 0$ loses stability when the eigenvalues of $\gamma(\mu) = \text{Re}\gamma(\mu) + i\,\text{Im}\gamma(\mu)$ of the spectral problem for $V \equiv 0$,

$$-\gamma\zeta + L_0\zeta + \mu L_1\zeta = 0, \tag{4}$$

cross the imaginary γ axis in conjugate pairs as μ passes through zero to the right,

$$\gamma(0) = i\omega_0, \qquad \overline{\gamma}(0) = -i\omega_0. \tag{5,6}$$

It is further assumed that $\gamma(\mu)$ is a simple isolated eigenvalue of L_0 and that the loss of stability is strict, $\text{Re}\gamma_\mu(0) < 0$.

The operators $L(\mu)$, L_0, L_1 and $N(\mu,V)$ are defined in a precise way by JS and will not be discussed here. In the analysis it is sufficient to think of the simplest realizations of (1) - the systems of ordinary differential equations considered by Hopf (1942). For ordinary differential equations, $V(t)$ is a vector, $L(\mu)$ is a matrix and $N(\mu;V)$ is the composition of matrices of functions of V and matrices independent of V. Our results hold for the general forms of L and N; the details of the computation in the demonstrations and the notations are more involved in the general case, but the results are the same. The extension of the results of this analysis to partial differential equations is immediate when $L(\mu)$ and $N(\mu;\cdot)$ satisfy the conditions stated by JS. For example, the results hold for nonlinear diffusion-reaction problems and for problems of fluid mechanics governed by the Navier-Stokes equations. Readers interested in this omitted aspect of the analysis may wish to consult Sattinger's monograph (1972).

To state the results, it is first necessary to specify the bifurcation problem and the spectral problem for the bifurcating solution. We introduce the scalar product

$$[a,b] = \frac{1}{2\pi}\int_0^{2\pi} a\cdot\bar{b}\ ds \qquad (7)$$

for complex-valued vectors $a(s)$, $b(s)$ which are 2π periodic in $s = \omega t$. The angle brackets designate volume-averaged integrals; the averaging is over the spatial region on which the vectors $a(x,s)$ and $b(x,s)$ are defined. For ordinary differential equations, $\langle a\cdot b\rangle = a\cdot b$. Real-valued bifurcating time-periodic solutions $u(s;\varepsilon)$ of (1), with L and N given by (2) and (3), satisfy

$$Ju + N(u,u) = 0, \qquad 2\varepsilon^2 = [u\cdot u], \qquad u(s) = u(s+2\pi) \qquad (8)$$

where

$$Ju = \omega(\varepsilon)\dot{u} + L_0 u + \mu(\varepsilon)L_1 u, \qquad \dot{u} \equiv \frac{du}{ds} \qquad (9)$$

and

$$\begin{pmatrix} u(s;\varepsilon) \\ \omega(\varepsilon)-\omega_0 \\ \mu(\varepsilon) \end{pmatrix} = \sum_{\ell=1}^{\infty} \varepsilon^{\ell} \begin{pmatrix} u_\ell(s) \\ \omega_\ell \\ \mu_\ell \end{pmatrix} \qquad (10)$$

are convergent power series in some complex neighborhood of $\varepsilon = 0$. The Taylor coefficients in (10) have the following properties:

$$\omega_0 = \mathrm{Im}\gamma(0),$$

$$\omega_{2\ell-1} = \mu_{2\ell-1} = 0 \qquad \ell \geqslant 1 \qquad (11)$$

and

$$u_1(s) = z_1 + z_2$$

where

$$z_1 = e^{-is}\zeta \text{ and } z_2 = \bar{z}_1 \qquad (12)$$

and ζ is the eigenfunction of L_0 belonging to the simple eigenvalue $i\omega_0$. The amplitude of ζ is fixed by the requirement that

$$[u_1\cdot u_1] = 2\left\langle|\zeta|^2\right\rangle = 2. \qquad (13)$$

The coefficients in the series (10) may be uniquely and sequentially determined from the boundary value problems which arise from (8) and (10). These problems are all in the form

$$J_0 u_\ell + f_\ell(s) = 0, \qquad J_0 = \omega_0\frac{d}{ds} + L_0 \qquad (14)$$

where u_ℓ and 2π-periodic functions satisfying a normalizing condition arising from (8). The Fredholm alternative for these problems is proved in lemmas of section 7 in JS. The perturbation problems are uniquely solvable and have bounded inverses when

$$[f_\ell\cdot z_1^*] = [f_\ell\cdot z_2^*] = 0$$

where $J_0^* z_1^* = J_0^* z_2^* = 0$ are eigenvalue problems for the adjoint operator J_0^* (see JS),

$$z_1^* = e^{-is}\zeta^*, \qquad z_2^* = \bar{z}_1^*$$

where

$$i\omega_0\zeta^* + L_0^*\zeta^* = 0 \tag{15}$$

and L_0^* is the adjoint operator for L_0. In the perturbation problem, $f_\ell(s)$ is real-valued and the one complex condition,

$$[f_\ell] \equiv [f_\ell \cdot z_1^*] = 0, \tag{16}$$

suffices for unique solvability. The amplitude of ζ^* is selected so that

$$\left.\begin{aligned} [u_1] &= \left\langle \zeta\cdot\zeta^* \right\rangle = 1. \\ [\dot{u}_1] &= -i. \end{aligned}\right\} \tag{17}$$

Then

The formula

$$-\gamma_\mu + [L_1 u_1] = 0, \qquad \gamma_\mu = \left.\frac{d\gamma}{d\mu}\right|_{\mu=0} \tag{18}$$

follows easily from (4), (15) and (17). The assumption that $V \equiv 0$ loses stability strictly as μ is increased past zero implies that $\mathrm{Re}\gamma_\mu < 0$.

The spectral problem for the conditional stability of (10) is obtained by introducing disturbances of the form

$$\begin{aligned} V &= \delta e^{-\sigma t}\Gamma + u(s;\varepsilon), \\ \Gamma &= \alpha(\varepsilon)\dot{u}(s;\varepsilon) + \gamma(s;\varepsilon) \end{aligned} \tag{19}$$

into (1) followed by linearization, $\delta \to 0$. The function Γ or, equivalently, the function γ, may be normalized by any convenient convention. We find that

$$\tau\dot{u} - \sigma\gamma + \mathcal{J}\gamma = 0 \tag{20}$$

where

$$\mathcal{J}(\cdot) = J(\cdot) + N(u,\cdot) + N(\cdot,u)$$

and

$$\tau = -\sigma\alpha.$$

According to Floquet theory, solutions of $\gamma(s)$ of (20) must be 2π-periodic functions of s. Moreover (see JS, section 5),

$$\begin{pmatrix} \gamma(s;\varepsilon)-u_1(s) \\ \tau(\varepsilon) \\ \sigma(\varepsilon) \end{pmatrix} = 2\begin{pmatrix} u_2(s) \\ \omega_2-\mu_2\mathrm{Im}\gamma_\mu \\ 0 \end{pmatrix}\varepsilon + \sum_{\ell=2}\varepsilon^\ell \begin{pmatrix} \gamma_\ell(s) \\ \tau_\ell \\ \sigma_\ell \end{pmatrix} \tag{21}$$

where τ_ℓ and σ_ℓ are real; and

$$i\tau_1 + \sigma_2 = 2(i\omega_2 - \mu_2\gamma_\mu). \tag{22}$$

The equation

$$\sigma_2 = -2\mu_2\mathrm{Re}\gamma_\mu, \qquad \mathrm{Re}\gamma_\mu < 0 \tag{23}$$

shows that subcritical solutions ($\mu_2 < 0$) are unstable. The series (21) has a finite, but possibly small, radius of convergence. The proof of convergence follows a slightly different path which allows the use of the implicit function theorem (see JS).

We are now ready to state and prove an extension of Hopf's theorem. The extension takes form as a factorization theorem. The factorization holds globally provided only that the quantities mentioned in the theorem are continuous functions of ε. No matter what the regularity properties of the solution may be for large values of ε they are regular analytic functions in some circle at the origin of the complex ε plane.

Suppose $\underset{\sim}{u}(\underset{\sim}{x},s;\varepsilon)$, $\omega(\varepsilon)$ and $\mu(\varepsilon)$ are real analytic functions on an open interval I_1 containing the point $\varepsilon = 0$. Then,

$$\underset{\sim}{\phi}(\underset{\sim}{x},s;\varepsilon) = \underset{\sim}{u}_\varepsilon(\underset{\sim}{x},s;\varepsilon) + \mu_\varepsilon(\varepsilon)\hat{\underset{\sim}{\phi}}(\underset{\sim}{x},s;\varepsilon),$$

$$\tau(\varepsilon) = \omega_\varepsilon(\varepsilon) + \mu_\varepsilon(\varepsilon)\hat{\tau}(\varepsilon) \tag{24}$$

and

$$\gamma(\varepsilon) = \mu_\varepsilon(\varepsilon)\hat{\gamma}(\varepsilon)$$

where $\hat{\underset{\sim}{\phi}}(\underset{\sim}{x},s;\varepsilon)$, $\hat{\tau}(\varepsilon)$ and $\hat{\gamma}(\varepsilon)$ are real analytic functions on an interval $I_2 < I_1$ containing the point $\varepsilon = 0$. Moreover, $\hat{\tau}(\varepsilon)$ and $\hat{\gamma}(\varepsilon)/\varepsilon$ are even functions of ε and such that

$$\hat{\gamma}_1 = -\text{re}\ \sigma_\mu, \qquad \tau_0 = -\text{im}\ \sigma_\mu.$$

The representation for $\sigma(\varepsilon)$ shows that $\sigma(\varepsilon)$ has all of the zeros of μ_ε. Unfortunately, we cannot assert that the function $\hat{\sigma}(\varepsilon)$ is of one sign when ε is large. When ε is small the sign of $\hat{\sigma}$ is known and the representation (24) leads to an extension of Hopf's stability theorem (see Theorem 2). It is of particular interest to determine the sign of $\hat{\sigma}(\varepsilon)$ at critical points $\tilde{\varepsilon}$, defined by

$$\mu^{\langle 1\rangle} = \left.\frac{d\mu}{d\varepsilon}\right|_{\varepsilon=\tilde{\varepsilon}} \equiv \mu_\varepsilon(\tilde{\varepsilon}) = 0 \tag{25}$$

of the bifurcating curve.

Proof of Theorem 1. We first introduce the representation (24) into (20) and find that

$$(\omega_\varepsilon+\mu_\varepsilon\hat{\tau})\dot{u} + \mathcal{J}u_\varepsilon + \mu_\varepsilon\mathcal{J}\hat{\gamma} - \mu_\varepsilon\hat{\sigma}(u_\varepsilon+\mu_\varepsilon\hat{\gamma}) = 0 \tag{26}$$

Differentiating (8) with respect to ε, we find that

$$\mathcal{J}u_\varepsilon + \omega_\varepsilon\dot{u} + \mu_\varepsilon L_1 u = 0. \tag{27}$$

When equation (27) is subtracted from (26), $\mathcal{J}u_\varepsilon + \omega_\varepsilon\dot{u}$ cancels and μ_ε may be factored from each of the remaining terms. It follows that

$$\mathcal{J}\hat{\gamma} + \hat{\tau}\dot{u} - L_1 u - \hat{\sigma}(u_\varepsilon+\mu_\varepsilon\hat{\gamma}) = 0. \tag{28}$$

The factorization (28) does not require analyticity in ε. To establish (24) under better hypotheses which replace analyticity with continuity, it would be necessary to prove the existence of continuous (in ε) functions $\hat{\underset{\sim}{\gamma}}(s;\varepsilon) = \hat{\gamma}(s+2\pi;\varepsilon)$, $\hat{\tau}(\varepsilon)$ and $\hat{\sigma}(\varepsilon)$ solving (28). I was unable to construct such an existence theorem. However, (28) does have a unique solution which is analytic in ε and has a power series converging in some circle centered at the origin of the complex ε plane. To prove this, we first assume that these functions have the following representations:

$$\begin{bmatrix} \hat{\gamma}(s;\varepsilon) \\ \hat{\tau}(\varepsilon)-\hat{\tau}_0 \\ \hat{\sigma}(\varepsilon) \end{bmatrix} = \sum_{\ell=1} \varepsilon^{\ell} \begin{bmatrix} \hat{\gamma}_{\ell}(s) \\ \hat{\tau}_{\ell} \\ \hat{\sigma}_{\ell} \end{bmatrix} \tag{29}$$

Inserting (29) and (10) into (28), we find that

$$J_0\hat{\gamma}_1 - L_1u_1 + \hat{\tau}_0\dot{u}_1 - \hat{\sigma}_1u_1 = 0, \tag{30}$$

$$J_0\hat{\gamma}_2 - L_1u_2 + \sum_{n+m=2}[\hat{\tau}_n\dot{u}_m-(m+1)u_{m+1}\hat{\sigma}_n+\mathcal{J}_n\hat{\gamma}_m] = 0, \tag{31}$$

and

$$\begin{aligned} J_0\hat{\gamma}_{\ell} - L_1u_{\ell} &+ \sum_{n+m=\ell}[\hat{\tau}_n\dot{u}_m-(m+1)u_{m+1}\hat{\sigma}_n+\mathcal{J}_n\hat{\gamma}_m] \\ &- \sum_{n+m+r=\ell}(n+1)\mu_{n+1}\hat{\sigma}_m\hat{\gamma}_r = 0, \qquad \ell \geqslant 3 \end{aligned} \tag{32}$$

where

$$\mathcal{J}_n(0) = \omega_n\frac{d}{ds}(\cdot) + \mu_nL_1(\cdot) + N(\cdot,u_n) + N(u_n,\cdot).$$

We next invoke the solvability condition (16) and use (17) to find that

$$\gamma_{\mu} + i\hat{\tau}_0 + \hat{\sigma}_1 = 0, \tag{33}$$

$$[L_1u_2] + i\hat{\tau}_1 + \hat{\sigma}_2 - \hat{\tau}_0[\dot{u}_2] + 2\hat{\sigma}_1[u_2] + [\mathcal{J}_1\hat{\gamma}_1] = 0 \tag{34}$$

and

$$\begin{aligned} [L_1u_{\ell}] + i\hat{\tau}_{\ell-1} + \hat{\sigma}_{\ell} &- \sum_{\substack{n+m=\ell \\ n\neq\ell}}\{\hat{\tau}_{n-1}[\dot{u}_{m+1}] - (m+1)\hat{\sigma}_n[u_{m+1}] \\ &+ [\mathcal{J}_n\hat{\gamma}_m]\} - \sum_{n+m+r=\ell}(n+1)\mu_{n+1}\hat{\sigma}_m[\hat{\gamma}_r], \qquad \ell \geqslant 3 \end{aligned}$$

The operator J_0 in equations (30), (31) and (32) has a bounded inverse on the complement of the null space of the operator J_0. When normalized in any convenient way, the solutions $\hat{\gamma}_{\ell}$ are unique. The normalized coefficients $\hat{\gamma}_{\ell}$, $\hat{\tau}_{\ell}$ and $\hat{\sigma}_{\ell}$ may be determined uniquely and sequentially and the series (29) converge when ε is sufficiently small. The proof of convergence copies the proof given in section 7 and 8 of JS. The functions $\hat{\gamma}(\varepsilon)$, $\hat{\tau}(\varepsilon)$ and $\hat{\sigma}(\varepsilon)$ may then be extended as real analytic functions defined on the interval I_2 of analyticity. In general, I_2 could not extend beyond the interval I_1 of analyticity of the operators defined in (20). This completes the proof of theorem 1.

When $\mu_2 \neq 0$, representation $\sigma = \mu_{\varepsilon}\hat{\sigma}$ leads to equation (23). This equation was derived by Hopf and used by him to prove the instability of subcritical solutions and the stability of supercritical solutions. The following local extension of Hopf's theorem may now be proved.

Theorem 2. When ε is small,

$$\sigma(\varepsilon) = -\mu_\varepsilon \varepsilon \mathrm{Re}\gamma_\mu + \mu_\varepsilon O(\varepsilon^3) \tag{35}$$

Subcritical solutions are unstable and supercritical solutions are stable.

Stability, in theorem 2 and throughout this paper, is in the sense of linearized theory. The result stated in theorem 2 does not require that $\mu_2 \neq 0$. To prove (35) we first note, following the computation (5.10) of JS, that $i\hat{\tau}_1 + \hat{\sigma}_2 = 0$. Hence, $\hat{\sigma}_2 = 0$ and (35) follows directly from (24) and (33).

The stability of steady bifurcating solutions can also be studied by the factorization method. The analysis follows along the lines laid out in the analysis of the stability of time-periodic bifurcating solutions. The following theorem holds:

Theorem 3 (Steady bifurcation). Suppose the $u(\varepsilon)$ and $\mu(\varepsilon)$ are real analytic functions on an open interval I_1 containing the point $\varepsilon = 0$. Then

$$\gamma(\varepsilon) = u_\varepsilon + \mu_\varepsilon \hat{\gamma}(\varepsilon)$$

and (36)

$$\sigma(\varepsilon) = \mu_\varepsilon \hat{\sigma}(\varepsilon)$$

where $\hat{\gamma}(\varepsilon)$ and $\hat{\sigma}(\varepsilon)$ are real analytic functions on an open interval I_2 containing the point $\varepsilon = 0$ and

$$\hat{\sigma}_1 = -\sigma_\mu + O(\varepsilon). \tag{37}$$

If $\mu_1 \neq 0$, then the bifurcation is two-sided and the subcritical branch is unstable.

If $\hat{\sigma}(\varepsilon)$ does not change sign, then $\sigma(\varepsilon)$ is negative when $\mu_\varepsilon > 0$ and is positive when $\mu_\varepsilon < 0$ (see Figs. 1 and 2). In the simplest situations, those in which $\hat{\sigma}$ controls stability and does not change sign, the unstable bifurcating solution regains stability as it turns around a critical point of the bifurcation curve. For partial differential equations and ordinary differential equations in R_n with $n > 2$ it is possible to have secondary instability and repeated branching. $\hat{\sigma}(\varepsilon)$ may exist but fail to control stability. For this reason, it is not possible to give a generally valid interpretation of the physical implications of the fundamental factorization. In examples of steady bifurcation in which it has been possible to construct global representations of the subcritical branch, $\hat{\sigma}(\varepsilon) \neq 0$ and stability is associated uniquely with the sign of μ_ε. In such cases we get snap-through instabilities. Computed global representations of subcritical bifur-

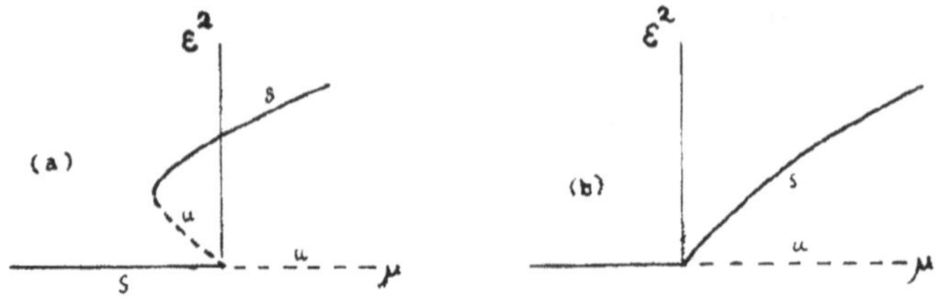

Fig. 1: Time-periodic bifurcation at a simple eigenvalue is one-sided. The time periodic solution bifurcates subcritically in (a) and supercritically in (b). Assuming $\hat{\sigma}$ is of one sign and controls stability, branches for which μ decreases as ε^2 increases are unstable.

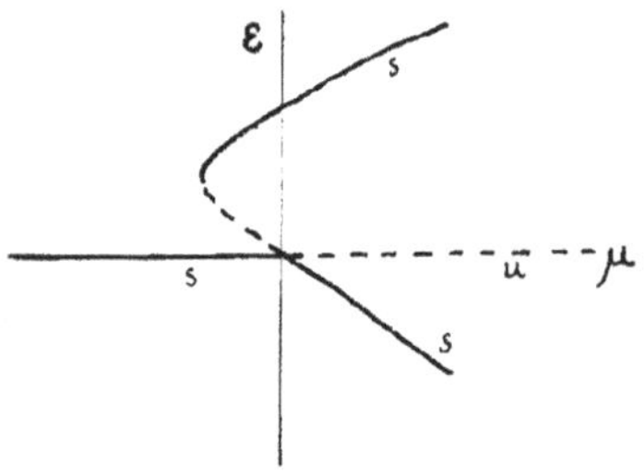

Fig. 2: Steady bifurcation at simple eigenvalue is usually two-sided. See caption for Fig. 1.

cation of time-periodic solutions are rare but, again, in one example, the numerical study of bifurcating time-periodic Poiseuille flow by Zahn, Toomre, Spiegel and Gough, (1973), $\hat{\sigma}(\varepsilon) \neq 0$ and we have restabilization of the subcritical branch and snap-through instabilities. It is necessary to add that, though the computations of Zahn, et al. proceed from a severely truncated version of the Navier-Stokes equation, the factorization (24) applies equally to the full equations and to the truncated version. Zahn, et al. consider traveling wave solutions of their equations; more general disturbances could possibly lead to instability and repeated branching on the conditionally stable upper branch of the bifurcation curve beyond the critical point. Assuming for the sake of the argument, that $\hat{\sigma}(\varepsilon)$ controls stability and that stability is associated uniquely with the sign of μ_ε, we are again led to a bifurcation picture for snap-through instability; at subcritical values of μ, $\mu_G < \mu < 0$, there are two conditionally stable solutions: laminar Poiseuille flow and time-periodic bifurcating Poiseuille flow on the stable subcritical upper branch of $\mu(\varepsilon)$, where $\mu_\varepsilon > 0$. The analysis applies to spatially periodic disturbances in infinitely long pipes and comparisons with experiments in pipes of finite length are at best suggestive. In finite pipes, where $\mu \in (\mu_G, 0)$, there also seem to be two "stable" solutions, one of which is laminar (Wygnanski and Champagne, 1973; Wygnanski, Sokolov, Friedman, 1975). The flow is spatially segregated into distinct patches of traveling packets of laminar and turbulent flow (turbulent "puffs" when μ is near μ_G, and "slugs" at higher values of μ). The transition from laminar to turbulent flow at a fixed place occurs suddenly as a puff or slug sweeps over the place, and the reverse transition occurs just as suddenly when it leaves the place. These observations suggest a sort of cycling in "phase space" between two distinct relatively stable but weakly attracting solutions.

This work was supported under the U.S. National Science Foundation grant GK 12500.

References

Iooss, G., Existence et stabilité des solutions périodiques de certains problèmes d'évolution du type Navier-Stokes. Arch. Rational Mech. Anal. 47, 301-329 (1972).

Hopf, E., Abzweigung einer periodischen Lösung eines Differentialsystems. Berichte der Mathematisch-Physikalischen Klasse der Sächsischen Akademie der Wissenschaften zu Leipzig, XCIV, 1-22 (1942).

Joseph, D. D. & D. A. Nield, Stability of bifurcating time-periodic and steady solutions of arbitrary amplitude. Arch. Rational Mech. Anal. (forthcoming)

Joseph, D. D., & D. H. Sattinger, Bifurcating time periodic solutions and their stability. Arch. Rational Mech. Anal. 45, 79-109 (1972).

Marsden, J. The Hopf bifurcation for nonlinear semigroups. Bull. Am. Math. Soc. 79, 3, 537-541 (1973).

Sattinger, D. H., Topics in Stability and Bifurcation Theory. Lecture Notes in Mathematics, Vol. 309. Berlin, Heidelberg, New York: Springer: 1973.

Wygnanski, I. J. & F. H. Champagne. On transition in a pipe. Part 1. The origin of puffs and slugs and the flow in a turbulent slug. J. Fluid Mech. 59, 281-335 (1973).

Wygnanski, I. J., M. Sokolov & D. Friedman. On transition in a pipe. Part 2. The equilibrium puff. J. Fluid Mech. 69, 283-305 (1975).

Zahn, J. P., J. Toomre, E. A. Spiegel & D. O. Gough, Nonlinear cellular motions in Poiseuille channel flow. J. Fluid Mech. 64, 319-45 (1974).

MESURES ET DIMENSIONS

par J.-P. Kahane

1. Mesure de Hausdorff.

Soit h une fonction croissante appliquant $\mathbb{R}^+$ sur $\mathbb{R}^+$. Soit E un espace métrique, A une partie de E. La mesure de Hausdorff de A relativement à h est

$$\mu_h(A) = \lim_{\delta \to 0} \mu_{h,\delta}(A)$$

où

$$\mu_{h,\delta}(A) = \inf \sum_i h\ (\text{diamètre } B_i)$$

la borne inférieure étant prise pour tous les recouvrements de A par des boules B_i de diamètre $\leq \delta$. Noter que $0 \leq \mu_{h,\delta}(A) \leq \infty$ et que $\mu_{h,\delta}(A)$ augmente quand δ diminue. Ainsi $0 \leq \mu_h(A) \leq \infty$.

μ_h est une mesure extérieure, c'est-à-dire que $\mu_h(\emptyset) = 0$ et $\mu_h(\bigcup_{n=1}^{\infty} A_n) \leq \sum_{n=1}^{\infty} \mu_h(A_n)$. Les ensembles μ_h-mesurables sont les parties A de E telles que les inclusions $B \subset A$ et $C \subset E \setminus A$ impliquent $\mu_h(B \cup C) = \mu_h(B) + \mu_h(C)$. Ils forment une tribu, et μ_h est une mesure au sens usuel (totalement additive) sur cette tribu. Tous les boréliens de E sont μ_h-mesurables.

Si E est $\mathbb{R}^n$, muni de la norme $\|x\| = \sup |x_j|$, et si $h(t) = t^n$, μ_h est la mesure extérieure de Lebesgue.

On va se borner désormais au cas où E est $\mathbb{R}^n$ normé, sans spécifier la norme (un changement de norme change μ_h, mais c'est généralement sans importance). On va se borner aussi au cas où $h(2t) = O(h(t))$ $(t \to 0)$.

Lemme de Frostman. Soit K un compact dans $\mathbb{R}^n$. On a $\mu_h(K) > 0$ si et seulement si K porte une mesure de probabilité σ telle que, pour toute boule B, on ait

$$\sigma(B) \leq C\, h\,(\text{diamètre } B)$$

C étant un réel ne dépendant que de σ.

Si $h(t) = t^\alpha$, μ_h s'appelle la mesure de Hausdorff en dimension α. On peut la noter μ_α. Observons l'effet d'une dilatation sur la mesure :

$$\mu_\alpha(\lambda A) = \lambda^\alpha \mu_\alpha(A).$$

Exemples.

1. (Hausdorff). Supposons h strictement concave, et, pour simplifier, $h(1)=1$. Posons $h(2^{-n}) = \xi_1 \xi_2 \ldots \xi_n$ (donc $0 < \xi_n < \frac{1}{2}$). L'ensemble parfait symétrique $E = E(\xi_1, \xi_2, \ldots)$, construit à la manière de l'ensemble triadique de Cantor, mais en prenant ξ_n pour rapport de dissection à la n-ième étape, satisfait $\mu_h(E) = 1$.

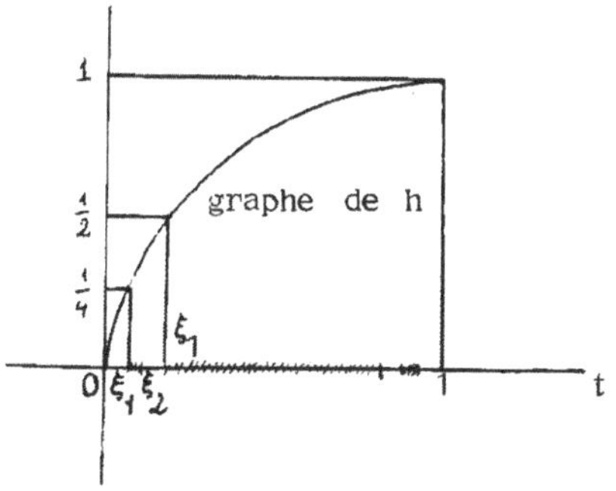

Pour $h(t) = t^\alpha$, les rapports ξ_n sont constants :

$$\xi_n = \xi = 2^{-\alpha}.$$

En particulier, pour $\alpha = \frac{\log 2}{\log 3}$, $\xi = \frac{1}{3}$ et E est l'ensemble triadique de Cantor.

Exercice : vérifier l'effet d'une homothétie de rapport ξ.

2. Courbes de Von Koch-Paul Lévy. On part d'un triangle isocèle T (plein) d'angle au sommet $\varphi > \frac{\pi}{2}$. On le dissèque en laissant tomber un triangle isocèle de même axe et d'angle au sommet $2\varphi-\pi$, de façon que restent deux triangles T_o et T_1, semblables à T, ayant pour bases les côtés de T. On répète cette dissection sur T_o et T_1, et ainsi de suite.

A la n-ième étape, on a 2^n triangles formant un ensemble fermé E_n. L'intersection des E_n, paramétrée de la façon naturelle, est une courbe K_φ. Pour $\varphi = \frac{2\pi}{3}$, c'est la courbe de Von Koch. Le cas limite $\varphi = \frac{\pi}{2}$ donne le courbe de Peano-Paul Lévy. Pour $\alpha = \frac{\log 2}{\log(2 \sin \frac{\varphi}{2})}$, on a $0 < \mu_\alpha(K_\varphi) < \infty$. Observer l'effet des similitudes de rapport $2 \sin \frac{\varphi}{2}$.

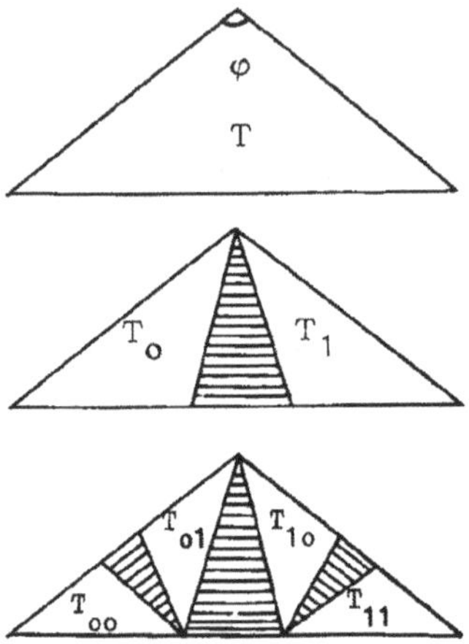

Voir ci-dessous une application de la courbe de Peano-Von Koch.

2. Dimension de Hausdorff.

Considérons deux fonctions h, soit h_1 et h_2, telles que $h_2(t) = o(h_1(t))$ $(t \to 0)$. Si $\mu_{h_1}(A) < \infty$, on a $\mu_{h_2}(A) = 0$. En conséquence

$$\inf \{\alpha \mid \mu_\alpha(A) = 0\} = \sup \{\alpha \mid \mu_\alpha(A) = \infty\}.$$

La veleur commune est la dimension de Hausdorff de A, notée $\dim A$.

Si $0 < \mu_\alpha(A) < \infty$, α est la dimension de A. D'où la dimension de l'ensemble de Cantor, des courbes de Von Koch, etc...

D'après le lemme de Frostman

$$\dim(K_1 \times K_2) \leq \dim K_1 + \dim K_2$$

(K_1 compact $\subset \mathbb{R}^n$, K_2 compact $\subset \mathbb{R}^m$, $K_1 \; K_2 \subset \mathbb{R}^{n+m}$). Dans les cas usuels, on a égalité ; par exemple, le carré de l'ensemble de Cantor a pour dimension $2\,\frac{\log 2}{\log 3}$.

Les variétés différentiables ont pour dimension de Hausdorff leur dimension ordinaire.

Si f est une application hölderienne d'ordre β, on a

$$\dim f(A) \leq \beta^{-1} \dim A .$$

Quand f est le paramétrage d'une courbe de Von Koch ou de Peano, on a l'égalité. C'est vrai aussi, presque sûrement, quand f est la fonction du mouvement brownien n-dimensionel pour $n \geq 2$; alors $\beta = \frac{1}{2}$ (R. Kaufman ; la partie $\forall A$ p. s. est bien connue ; mais p. s. $\forall A$ est plus difficile).

Les probabilités donnent beaucoup d'exemples d'ensembles de dimension non entière (images des processus de Lévy, zéros de fonctions aléatoires, supports de mesures aléatoires,...)

Autre exemple (Eggleston) : les nombres réels dont le développement décimal contient le chiffre j avec une fréquence p_j forment un ensemble de dimension

$$\sum_{o}^{g} p_j \log_{10} \frac{1}{p_j} .$$

Voici quelques procédés utiles pour la majoration ou le calcul de la dimension.

1 (Besicovitch et Taylor). Soit K un compact de mesure de Lebesgue nulle dans $[0,1]$, et I_j $(j = 1, 2, \dots)$ les intervalles contigus à K sur $[0,1]$. Si $\Sigma I_j^\alpha < \infty$, on a $\dim K \leq \alpha$.

2 (variante de 1). Soit K un compact de mesure de Lebesgue nulle dans $[0,1]^n$, et I_j les pavés dyadiques maximaux disjoints de K. Si Σ (diamètre $I_j)^\alpha < \infty$, on a $\dim K \leq \alpha$.

(Visiblement 2 équivaut à 1 pour $n = 1$; le transfert à $n > 1$ se fait par une courbe de Peano).

3 (Billinsley). Soit μ une mesure de probabilité sur $[0,1]^n$, et pour tout x à coordonnées irrationnelles soit $I_j(x)$ le pavé dyadique d'ordre j contenant x. Si on a

$$\lim_{j\to\infty} \frac{\log \mu(I_j(x))}{j \log 2} = -\alpha$$

pour μ-presque tout x, alors tout borélien de dimension $< \alpha$ est de μ-mesure nulle, et il existe un borélien de dimension α et de μ-mesure pleine. (Exercice : déduire cela du lemme de Frostman).

4 (Frostman). C'est l'identité de la dimension de Hausdorff et de la dimension capacitaire pour les compacts dans $\mathbb{R}^n$, qu'on va définir.

3. Dimension capacitaire, dimension de Fourier.

Soit K un compact dans $\mathbb{R}^n$. Désignons par $M_1^+(K)$ l'ensemble des mesures de probabilité sur K. On note $\hat{\sigma}$ la transformée de Fourier de σ, $\int \varphi(u)$ du l'intégrale de Lebesgue de φ dans $\mathbb{R}^n$ privé de la boule unité, et α un réel $\in \,]0,n[$.

Dire que K est de α-capacité non nulle, c'est dire qu'il existe $\sigma \in M_1^+(K)$ telle que

$$(*) \qquad \int |\hat{\sigma}(u)|^2 \, |u|^{\alpha-n} \, du < \infty$$

(théorie du potentiel). La borne supérieure des $\alpha \in \,]0,n[$ tels qu'il en soit ainsi est la dimension capacitaire de K (définition de Polya). C'est aussi sa dimension de Hausdorff (Frostman).

Même si α est voisin de n, (*) n'entraîne pas que $\hat{\sigma}(u)$ tende vers 0 quand $|u| \to \infty$. Certains ensembles, à définition géométrique simple et de dimension > 0 ne portent aucune mesure ni distribution $\neq 0$ dont la transformée de Fourier tende vers 0 à l'infini : ainsi la frontière d'un cube dans $\mathbf{R}^n$, toute partie compacte d'une variété linéaire de dimension n-1 dans $\mathbf{R}^n$, l'ensemble triadique de Cantor sur $\mathbf{R}$. Il existe dans $\mathbf{R}^n$ des ensembles compacts de dimension n ayant cette propriété.

Par contre, si $\hat{\sigma}(u)$ tend vers zéro à l'infini assez vite (p. ex. $\hat{\sigma}(u) = 0(|u|^{-\frac{\alpha}{2}-\varepsilon})$, $\varepsilon > 0$), on a (*). Appelons dimension de Fourier de K, et notons $\mathfrak{F}$-dim K, la borne supérieure des $\alpha \in \,]0,n[$ tels qu'il existe $\sigma \in M_1^+(K)$ avec

$$(**) \qquad \hat{\sigma}(u) = 0(|u|^{-\frac{\alpha}{2}}) \quad (|u| \to \infty).$$

On a $\mathfrak{F}$-dim K $\leq$ dim K. On vient de voir des exemples où l'inégalité est stricte.

Pour une sphère K dans $\mathbf{R}^n$, on a $\mathfrak{F}$-dim K = dim K = n-1. Sur $\mathbf{R}$, on ne connait aucune construction simple d'un K tel que $0 <$ $\mathfrak{F}$-dim K = dim K = $\alpha < 1$. Salem a montré que de tels ensembles existent pour tout $\alpha \in \,]0,1[$. Un énoncé métamathématique est qu'un compact aléatoire K satisfait très souvent

$\mathcal{F}$-dim K = dim K ; il en est ainsi pour les images d'un compact fixé arbitraire 1) par le mouvement brownien 2) par un processus stationnaire à accroissements indépendants 3) par un processus gaussien stationnaire, etc... etc...

La condition (**) impose beaucoup plus à K que la condition (*). Par exemple, (**) impose que toute projection de K sur une variété linéaire V ait une dimension $\geq \alpha$. Egalement (**) impose que $K + K + \ldots + K$ (addition algébrique) contienne un ouvert si le nombre de termes de la somme est assez grand. (*) n'exige rien de tel.

La dimension et la dimension de Fourier ont un certain rapport avec les théorèmes de trace. Détaillons un peu la chose.

Soit $\mathcal{E}$ un espace de distributions sur R^n, contenant $\mathcal{D}$ topologiquement, et K un compact dans R^n. Voici les premiers problèmes de trace qu'on peut se poser :

1) existe-t-il $\tau \in \mathcal{D}'(K)$, $\tau \neq 0$, tel que l'application $f \to f\tau$, définie sur $\mathcal{D}$, se prolonge par continuité de $\mathcal{E}$ dans $\mathcal{D}'(K)$?

2) existe-t-il $\sigma \in M_1^+(K)$, telle que l'application $f \to f\sigma$, définie sur $\mathcal{D}$, se prolonge par continuité de $\mathcal{E}$ dans $M(K)$, l'espace des mesures de Radon sur K ?

3) s'il en est ainsi, pour quelles valeurs de $p \geq 1$ l'application $f \to f|_K$, définie sur $\mathcal{D}$, se prolonge-t-elle par continuité de $\mathcal{E}$ dans $L^p(\sigma)$?

Supposons - c'est le cas pour beaucoup d'espaces usuels - que $\mathcal{E}$ soit un module sur $\mathcal{D}$. Alors 1) et 2) s'énoncent ainsi

1') existe-t-il $\tau \in \mathcal{E}' \cap \mathcal{D}'(K)$, $\tau \neq 0$?

2') existe-t-il $\sigma \in \mathcal{E}' \cap M_1^+(K)$?

Supposons de plus que $\mathcal{D}$ soit dense dans $\mathcal{E}$, on a alors une réponse partielle à 3) : l'application $f \to f|_K$ est prolongeable de $\mathcal{E}$ dans $L^1(\sigma)$.

Exemples : a) $\mathcal{E} = H^s = \left\{ f \mid \int |\hat{f}(u)|^2 |u|^{2s} du < \infty \right\}$, avec $0 < s \leq \frac{n}{2}$. La réponse à 2 est positive si et seulement si la capacité d'ordre $n - 2s$ de K est > 0. En particulier, la réponse est positive si $\dim K > n - 2s$, et négative si $\dim K < n - 2s$. La réponse à 1 et la réponse à 2 sont les mêmes dans certains cas (théorie du potentiel), et peut-être dans tous les cas (question ouverte).

b) $\mathcal{E} = \left\{ f \mid \int |\hat{f}(u)|\, |u|^{-\frac{\alpha}{2}} du < \infty \right\}$ avec $0 < \alpha < n$ (rappelons que $\int = \int_{\mathbb{R}^n \setminus \text{boule unité}}$; une définition plus explicite est la suivante : $\mathcal{E}$ est l'espace des cotransformées de Fourier des distributions $\in \mathcal{S}'$ qui coïncident avec une fonction sommable pour la mesure $|u|^{-\frac{\alpha}{2}} du$ hors de la boule unité). La réponse à 2 est positive si $\mathcal{F}$-dim $K > \alpha$, et négative si $\mathcal{F}$-dim $K < \alpha$. Il en est peut être de même pour la réponse à 1 (cela ne signifie pas que les réponses à 1 et à 2 soient les mêmes pour tous les K ; je n'en sais rien, mais j'en doute).

c) (Fefferman et Stein) $n = 2$, $\mathcal{E} = L^q(\mathbb{R}^2)$, K = cercle unité. Les réponses à 1 et à 2 sont positives si $1 \leq q < \frac{4}{3}$, négatives si $q \geq \frac{4}{3}$. On peut prendre pour σ la mesure équidistribuée sur K si $1 \leq q < \frac{4}{3}$, et la réponse à 3 est alors $1 \leq p \leq \frac{q}{3(q-1)}$.

4. Une application amusante de la courbe de Peano-Paul Lévy (Kakutani, communication orale, août 1974).

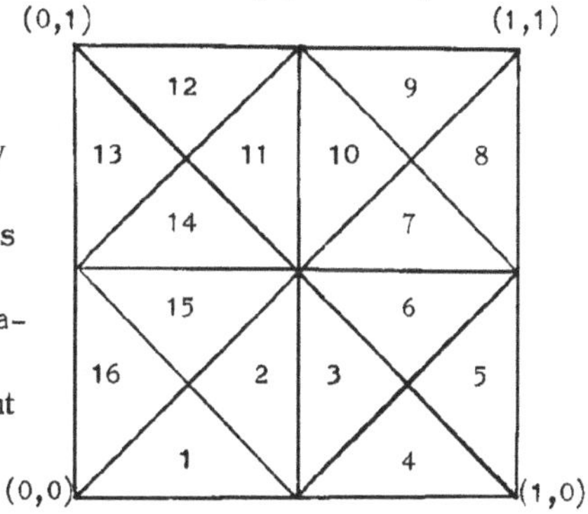

Construisons la courbe de Peano-Paul Lévy qui remplit le triangle $(0,0)$, $(1,0)$, $(\frac{1}{2},\frac{1}{2})$, et les trois courbes obtenues à partir d'elle par des rotations de $\frac{\pi}{2}$, π, $\frac{3\pi}{2}$ autour de $(\frac{1}{2}, \frac{1}{2})$. On obtient une courbe remplissant le carré $[0,1] \times [0,1]$, qu'on peut paramétrer par T de façon que l'image de la mesure de Lebesgue sur T soit la mesure superficielle sur le carré. Ainsi, le groupe des translations sur T induit sur le carré un groupe de transformations préservant la mesure

On peut vérifier que le carré de la distance des images de t et de t' ne dépasse pas $4|t-t'|$. Application : on donne n points dans le carré. On peut les ordonner sous la forme $M_1, M_2, \ldots M_n = M_o$, de façon que $\sum_1^n M_j M_{j+1}^2 \leq 4$. [Il suffit de les ordonner selon l'ordre du paramètre sur **T**].

5. Bibliographie.

C. A. Rogers, Hausdorff measures, Cambridge U. P. 1970 (on y trouve la théorie de la mesure extérieure, la justification du début du § 1, des exemples, une liste de références).

F. Hausdorff, Dimension und äusseres Mass, Math. Annalen 79 (1919) 157-179 (c'est la référence de base ; outre la théorie de la mesure et la définition de la dimension "fractionnaire", on y trouve l'exemple 1 du § 1).

O. Frostman, Potentiel d'équilibre et capacité des ensembles, Thèse, Lund 1935 (c'est un classique, malheureusement difficile à trouver, de la théorie du potentiel ; le "lemme de Frostman", § 1, intervient dans la démonstration de l'égalité de la dimension de Hausdorff et de la dimension capacitaire).

J.-P. Kahane et R. Salem, Ensembles parfaits et séries trigonométriques (les chapitres II et III sont tirés de Hausdorff, Frostman et Beurling ; l'exposé est fait sur R ou T ; c'est la référence la plus courante pour le lemme de Frostman ; le chapitre VIII contient les résultats de Salem relatifs à $\mathfrak{F}$-dim = dim (§ 3)).

Paul Lévy, Quelques aspects de la pensée d'un mathématicien, Paris 1970 (sur les courbes de Von Koch ; bouquin délectable, sans autre rapport avec le sujet).

J. M. Marstrand, R. Kaufman, articles cités par Rogers (dimension d'un produit cartésien, dimension des projections, etc...).

R. Kaufman, Comptes Rendus, 268 (1969), p. 727. (le mouvement brownien double la dimension).

B. Mandelbrot, oeuvres complètes, en particulier un bouquin à paraître (sur des exemples théoriques ou concrets d'ensembles de dimension non entière ; voir aussi Jacques Peyrière, Turbulence et dimension de Hausdorff, Comptes Rendus, t. 278 (1974)).

A. S. Besicovitch et S. J. Taylor, On the complementary intervals ..., J. London math. Soc. 29 (1954) 449-59 (sur l'énoncé 1 du § 2).

H. G. Eggleston, articles cités par Rogers (en particulier sur l'exemple donné au § 2).

P. Billingsley, Ergodic theory and information, Wiley 1965 (contient l'exemple d'Eggleston, et, sous une forme un peu plus générale, l'énoncé 3 du § 2).

J.-P. Kahane, Some random series of functions, Heath 1967 (mouvement brownien et $\mathfrak{F}$-dimension ; voir aussi Images browniennes des ensembles parfaits, Comptes Rendus t. 263 (1966) 613-615, et autres papiers du même).

C. Fefferman, rapport au congrès de Vancouver (1974)(contient les références, compléments et commentaires autour du théorème de trace c, § 3).

Jean Perrin, Atomes, Paris 1913 (c'est un des rares livres que N. Wiener aimait à citer ; l'introduction contient de belles choses sur le rôle putatif des ensembles singuliers en physique).

SINGULAR PERTURBATION AND SEMIGROUP THEORY

Tosio Kato

1. Introduction.

In this paper I consider the relationship between the problems of singular perturbation and semigroup theory for linear operators, with emphasis on the case when the unperturbed operator is "hyperbolic". The paper contains a small number of theorems, together with many remarks (which are often propositions loosely stated) and some examples.

In general, the problem of singular perturbation may be regarded as the problem of the resolvent convergence

(1.1) $$A_\varepsilon \xrightarrow[r]{} A_0 , \quad \varepsilon \downarrow 0 ,$$

for a family $\{A_\varepsilon ; 0 < \varepsilon < \varepsilon_0\}$ of linear operators. [In what follows all convergence refers to $\varepsilon \downarrow 0$, unless otherwise stated]. [In this paper I do not consider the problem of asymptotic expansion, which refines (1.1) by studying the rate of convergence].

I am mostly concerned with the case in which A_ε and A_0 are the negative generators of equi-bounded C_0-semigroups in a Banach space X (so that $\|e^{-tA_\varepsilon}\| \leq M$). Then (1.1) means that

(1.2) $$(A_\varepsilon+\lambda)^{-1} \xrightarrow[s]{} (A_0+\lambda)^{-1} , \quad \lambda>0 ,$$

or, equivalently, for sufficiently large λ, where $\xrightarrow[s]{}$ denotes strong convergence.

(1.2) is equivalent to the following : for any $f_0 \in X$,

(1.3) $$(A_\varepsilon+\lambda)u_\varepsilon = f_\varepsilon , \quad (A_0+\lambda)u_0 = f_0 , \text{ and } f_\varepsilon \rightarrow f_0$$

together imply $u_\varepsilon \rightarrow u_0$ (stationary problem).

According to the so-called Trotter-Kato theorem, however, (1.2) is also equivalent to

(1.4) $$e^{-tA_\varepsilon} \xrightarrow[s]{} e^{-tA_0} , \text{ uniformly on } t \in [0,T] ,$$

for any positive number T. In this sense (1.2) is equivalent to the following :

$$(1.5)\qquad \frac{du_\varepsilon}{dt} + A_\varepsilon u_\varepsilon = f_\varepsilon(t) \; , \quad \frac{du_o}{dt} + A_o u_o = f_o(t) \; ,$$

$f_\varepsilon(t) \longrightarrow f_o(t)$ boundedly on $[0,T]$, say,

and $u_\varepsilon(0) \longrightarrow u_o(0)$, together imply that

$u_\varepsilon(t) \longrightarrow u_o(t)$ uniformly on $[0,T]$ (nonstationary problem).

The situation is more complicated if one considers temporally inhomogeneous equations of evolution

$$(1.6)\qquad \frac{du_\varepsilon}{dt} + A_\varepsilon(t)u_\varepsilon = f_\varepsilon(t) \; , \quad \frac{du_o}{dt} + A_o(t)u_o = f_o(t), \text{ etc...}$$

since the time-dependence of $A_\varepsilon(t)$ etc. poses a new problem.

Remark 1. It is embarrassing that there is no satisfactory theorem for the convergence $u_\varepsilon(t) \longrightarrow u_o(t)$ to take place in (1.6), which reduces to the Trotter-Kato theorem when $A_\varepsilon(t)$ is independent of t . I did give such a theorem in [1], but it does not seem to be sufficiently strong in applications involving "hyperbolic" equations.

It is convenient to express the resolvent convergence (1.1) in a form slightly different from (1.2). (1.1) is equivalent to the following condition : for each u in a core for A_o , there exists a family $v_\varepsilon \in D(A_\varepsilon)$ such that

$$(1.7)\qquad v_\varepsilon \longrightarrow u \; , \quad A_\varepsilon v_\varepsilon \longrightarrow A_o u \; .$$

This condition is convenient in applications since the choice of v_ε is rather flexible.

2. Classification of singular perturbations.

There are different degrees of singularity among singular perturbations. To discuss this situation, I restrict myself in what follows to the case in which

$$(2.1)\qquad A_\varepsilon = A_o + \varepsilon S \; , \quad \varepsilon>0 \; , \quad D(A_\varepsilon) = D(S) \subset D(A_o) \; ,$$

with the standing assumption that the e^{-tA_ε} are equi-bounded semigroups.

When applying the criterion (1.7) to this case, it is convenient to distinguish between two cases.

Case I. For each u in a core for A_o , there exists a family $v_\varepsilon \in D(S)$ such that

$$(2.2) \qquad v_\varepsilon \to u\ , \quad A_0 v_\varepsilon \to A_0 u\ , \quad \varepsilon S v_\varepsilon \to 0\ .$$

Case II. This is not true.

Of course Case I is a milder case ; there may be a "boundary layer" involved, but it is relatively weak. Each case may be further classified into several cases with different degrees of singularity, as we shall discuss in the sequel.

3. Case I.

Case I occurs if and only if $D(S)$ is a core for A_0 .

Indeed, the first two relations in (2.2) already imply that $D(S)$ is a core for A_0 . If, conversely, $D(S)$ is a core for A_0 , we may simply set $v_\varepsilon = u$ for $u \in D(S)$ to satisfy (2.2).

Thus we recover the well-known result that $A_\varepsilon = A_0 + \varepsilon S \underset{r}{\longrightarrow} A_0$ if $D(S)$ is a core for A_0 .

A rather trivial example of Case I is given by the case in which A_0 is bounded.

Even though Case I is an easy case as a singular perturbation, it is in general not an easy task to prove that $D(S)$ is a core for a given operator A_0 (even if this is true). Such a proof is analogous to the proof that certain formally selfadjoint differential operators are essentially selfadjoint in a given Hilbert space, which is usually a nontrivial problem. There are many theorems dealing with this type of questions, but many of them are concerned with perturbed operators when it is assumed that the set in question is a core for the unperturbed operator. Theorems of this kind are not very useful for our purpose.

In this connection I state a theorem which is in fact useful. It is a generalization of a theorem due to Okazawa [2].

THEOREM 2. Let A and S be accretive operators in a real Hilbert space H , with S m-accretive, such that $D(A) \supset D(S)$. Assume that

$$(3.1) \qquad (Au, Su) \geqslant -M\|u\|\|Su\| \quad \text{for} \quad u \in D(S)\ .$$

Then (a) A is essentially m-accretive (that is, the closure A_0 of A is m-accretive). (b) $D(S)$ is a core for A_0 . (c) $A_\varepsilon = A + \varepsilon S$ is m-accretive for $\varepsilon > 0$ (with domain $D(S)$). (d) $A_\varepsilon \underset{r}{\longrightarrow} A_0$.

This theorem has a nontrivial application to the following example.

Example 3. (Linearized Navier-Stokes and Euler equations).

Let $\mathcal{O} \subset R^3$ be a bounded domain with smooth boundary, and let H be the subspace of the real Hilbert space $L^2(\mathcal{O})^3$ consisting of solenoidal vectors u ($\operatorname{div} u = 0$, $u_n = 0$ on $\partial\mathcal{O}$) . Let P be the orthogonal projection of $L^2(O)^3$ onto H . Let

$$(3.2) \qquad A = P(a.\mathrm{grad}) \; , \qquad D(A) = H \cap H^1(\mathcal{O})^3 \; ,$$

where a is a smooth function belonging to H and $H^1(\mathcal{O})$ denotes the Sobolev space of order 1 of L^2-type. Further let

$$(3.3) \qquad S = -P\Delta \; , \quad D(S) = H \cap H^2(\mathcal{O})^3 \cap H^1_o(\mathcal{O})^3 \; .$$

Then S is a positive selfadjoint operator in H (see Cattabriga [3], Ladyzhenskaya [4]). It is not difficult to show that (3.1) is satisfied if a is sufficiently smooth (e.g. $a \in C^1(\mathcal{O})^3 \cap H$) . It follows from Theorem 2 that A is essentially m-accretive and $A_\varepsilon = A + \varepsilon S \underset{r}{\longrightarrow} A_o$ (closure of A), which implies that (1.3) and (1.5) hold (convergence for vanishing viscosity).

The following theorem, which is a special case of Theorem 2, is useful in many applications.

__THEOREM 4.__ In Theorem 2 assume that S is nonnegative selfadjoint with $D(S^{1/2}) \subset D(A)$. Assume further that (3.1) holds with $M = 0$. Then $D(S^{1/2})$ is A_o-admissible, with the semigroup $\{e^{-tA_o}\}$ contractive on $D(S^{1/2})$ (see Kato [5]).

__Remark 5.__ The conclusion of the theorem means that e^{-tA_o} maps $D(S^{1/2})$ into itself and forms a contractive semigroup (with respect to the graph norm $u \mapsto (\|u\|^2 + \|S^{1/2}u\|^2)^{1/2}$). It implies also that $(A_o+\lambda)^{-1}$ maps $D(S^{1/2})$ into itself if $\lambda > 0$. [The family $A_\varepsilon = A + \varepsilon S$ is rather well-behaved in this case, which I want to call __Case__ I_o].

__Remark 6.__ The preceding remark contains a __regularity theorem__, in the sense that $(A_o+\lambda)u = f \in D(S^{1/2})$ implies $u \in D(S^{1/2})$. Indeed, in most applications S is an elliptic operator so that $D(S^{1/2})$ can be described explicitly.

__Example 7.__ As an example for Remarks 5 and 6, consider the operator A of Example 3, together with the operator S_1 defined as follows. S_1 is by definition the selfadjoint operator in H associated with the quadratic form $u \mapsto \|\mathrm{grad}\, u\|^2$ with domain $H \cap H^1(\mathcal{O})^3$, which is also the domain of $S_1^{1/2}$. Formally $S_1 = -P\Delta$, with the boundary condition for $u \in D(S_1)$ that $\partial u/\partial n$ be normal to $\partial\mathcal{O}$ (in addition to $u_n = 0$).

It is not difficult to show that the conditions of Theorem 4 are satisfied for $A+\beta$ and S_1 , where β is a sufficiently large real number. It follows that one has the regularity result : $(A_o+\lambda)u = f \in H \cap H^1(\mathcal{O})^3$ implies $u \in H \cap H^1(\mathcal{O})^3$ if

$\lambda > \beta$. This is a useful information on A_o , in addition to those given by Example 3. Moreover, one can use the new information to prove the convergence result stated in Example 3 in a different way.

Example 8. Theorem 4 is also useful for deducing similar results for a first-order system

$$(3.4) \qquad A = \sum_{j=1}^{m} a_j(x)D_j + a_o(x) , \quad D_j = \partial/\partial x_j , \quad x \in \mathcal{O} ,$$

where $\mathcal{O}$ is a bounded domain in R^m with smooth boundary and the $a_j(x)$ are real $N \times N$ matrices depending smoothly on x . A acts on N-vector valued functions $u = (u_1(x),\ldots,u_N(x))$ defined on $\mathcal{O}$. We assume for simplicity that the $a_j(x)$ for $j \geq 1$ are symmetric matrices. For $x \in \partial\mathcal{O}$, we define $a_n(x) = \sum n_j(x)\, a_j(x)$, where the $n_j(x)$ are the components of the unit outer normal to $\partial\mathcal{O}$ at x . We regard A as an operator in $H = L^2(\mathcal{O})^N$ with domain appropriately restricted (see below).

Suppose now that $a_n(x) \geq 0$ for $x \in \partial\mathcal{O}$. Then it can be shown that the assumptions of Theorem 4 are satisfied for $A+\beta$ (for sufficiently large β) and $S_1 = -\Delta$, with the Neumann boundary condition. In this way we obtain the regularity result that $(A_o+\lambda)u = f \in H^1(\mathcal{O})^N = D(S_1^{1/2})$ implies $u \in H^1(\mathcal{O})^N$ if $\lambda > \beta$. Note that we are not assuming that $a_n(x)$ is strictly positive. Thus the result is not contained in the standard result (such as Tartakoff [6]).

If we assume, on the other hand, that $a_n(x) \leq 0$ for $x \in \partial\mathcal{O}$, the assumptions of Theorem 4 are satisfied for $A+\beta$ and $S = -\Delta$, with the Dirichlet boundary condition. It follows, as above, that $(A_o+\lambda)u = f \in H_o^1(\mathcal{O})^N = D(S^{1/2})$ implies $u \in H_o^1(\mathcal{O})^N$. This regularity result would appear somewhat incomplete inasmuch as H_o^1 appears instead of H^1 . But it is easy to remove this restriction by an auxiliary argument. Again note that we do not assume that $a_n(x)$ is strictly negative.

The general case in which $a_n(x)$ is neither nonnegative nor nonpositive can be handled by the same method, at least if $a_n(x)$ is nonsingular for $x \in \partial\mathcal{O}$. In this case, however, a simple choice $S = -\Delta$ will not work and the method is not altogether trivial.

Remark 9. One can prove a theorem, corresponding to Theorem 2, for the evolution equations (1.6), where we assume that $A_\varepsilon(t) = A(t) + \varepsilon S$ with $A(t)$ and S satisfying the conditions of Theorem 2. I shall not give here a precise formulation, however. If one wants to minimize the assumption on the dependence of $A(t)$ on t , one would have to assume that S is nonnegative selfadjoint. Then the perturbed equation is "parabolic" and one could establish the existence of the solution under a mild t-dependence of $A(t)$. The convergence $u_\varepsilon(t) \to u_o(t)$ could then be proved without difficulty, but the limit function u_o may not be a strong solution of the limit equation.

Remark 10. If one makes a stronger assumption that $A(t)$ and S satisfy the assumptions of Theorem 4, however, the situation becomes much better. Since $D(S^{1/2})$ is $A(t)$-admissible in this case (Theorem 4), the theory of evolution equations of "hyperbolic" type (see [5]) is available under a minimum assumption on the t-dependence of $A(t)$. Time-dependent problems corresponding to Example 3 (linearized Navier-Stokes and Euler equations) have been studied by this method and the convergence $u_\varepsilon(t) \to u_o(t)$ was proved by Lai [7], under the assumption that $t \mapsto a(t,x)$ is continuous from $[0,T]$ to $H \cap C^1(\mathcal{O})^3$. It should be noted that in this method one can first prove the existence and uniqueness for solutions of the Euler equation, depending on the result given in Example 7. [The convergence mentioned above is stated in [4] without proof].

Remark 11. These results on the linearized Navier-Stokes and Euler equations do not shed much light on the problem involving nonlinear Navier-Stokes equations, since high smoothness has to be assumed for $a(t,x)$. But they do show that the difficulty with the Navier-Stokes equation is essentially due to nonlinearity.

In this connection I note that in [7] Lai has shown that the L^2-convergence of the solution of the Navier-Stokes equation to that of the Euler equation takes place if one assumes that

$$\|\Delta u_\varepsilon(t)\| \leq \text{const}(\varepsilon t)^{-\alpha} \quad \text{for some} \quad \alpha < 3/2 ,$$

where $\| \ \|$ is the $L^2(\mathcal{O})^3$-norm, provided that the data are sufficiently smooth.

4. Case II.

This is a more difficult case, involving a genuine boundary layer. It appears that there is no known general theorem that can handle this case to prove the resolvent convergence (1.1). One has to construct a family $\{v_\varepsilon\}$ satisfying (1.7) explicitly.

If we write

$$w_\varepsilon = u - v_\varepsilon \in u - D(S) \qquad \text{("boundary layer")} \tag{4.1}$$

and use (2.1), (1.7) is satisfied if

$$w_\varepsilon \to 0 , \quad A_o w_\varepsilon + \varepsilon S' w_\varepsilon \to 0 , \tag{4.2}$$

where S' is an extension of S such that $u \in D(S')$. In general S' will depend on u , but one can usually find such an S' easily if u is "smooth" (see Example 12 below).

The difficulty in Case II is that in general $A_0 w_\varepsilon$ will not tend to zero even weakly ; otherwise u would be in the closure of A_0 restricted to $D(S)$, which cannot be the case for every u in a core for A_0 . Thus (4.2) requires that there should be a cancellation between $A_0 w_\varepsilon$ and $\varepsilon S' w_\varepsilon$.

Note, on the other hand, that <u>it suffices to construct a family</u> w_ε <u>satisfying</u> (4.2) for each u in core for A_0 .

Classical examples of Case II are found in Levinson [8], Ladyzhenskaya [9], Vishik and Lyusternik [10] ; see also Lions [11]. Here we shall illustrate the problem by considering a singular perturbation of a symmetric system of first order by an elliptic operator of second order, thus partially generalizing the results of these authors.

<u>Example 12</u>. Let $\mathcal{O} \subset R^m$ and the operator A be as in Example 8. Let

$$S = - \sum_{j,k=1}^{m} b_{jk}(x)\, D_j\, D_k + \sum_{j=1}^{m} b_j(x)\, D_j + b(x) \ , \tag{4.3}$$

where the coefficients b_{jk} etc. are $N \times N$ real matrix-valued smooth functions defined on $\mathcal{O}$. We assume that $b_{jk}(x)^* = b_{jk}(x)$ and that S is strongly elliptic in the sense that

$$\sum \xi_j\, \xi_k\, b_{jk}(x) > 0 \quad \text{for} \quad 0 \neq \xi \in R^m \ , \quad x \in \mathcal{O} \ ; \tag{4.4}$$

note that the matrix in (4.4) is symmetric. We also assume that $\partial\mathcal{O}$ is not characteristic for A , which means that the matrix $a_n(x)$ defined in Example 8 is nonsingular for $x \in \partial\mathcal{O}$.

A and S act on vector-valued functions $u = \{u_1(x), \ldots, u_N(x)\}$ defined on $\mathcal{O}$. We regard these operators as linear operators in $H = L^2(\mathcal{O})^N$. In this sense A is formally skew symmetric and S is formally selfadjoint, each modulo an operator of lower order. Thus

$$A_\varepsilon = A + \varepsilon S \quad \text{with domain} \quad D(S) = H^2(\mathcal{O})^N \cap H_0^1(\mathcal{O})^N \tag{4.5}$$

is m-accritive for $\varepsilon > 0$. The question is whether there is an m-accretive operator A_0 such that $A_\varepsilon \underset{r}{\longrightarrow} A_0$.

We shall show that such an A_0 does exist ; it is a restriction of the formal differential operator A with a certain boundary condition. An interesting problem is to determine this boundary condition.

As is well known (see e.g. Lax-Phillips [12]), a boundary condition that makes A m-acrretive is given by

$$u(x) \in M(x) , \quad x \in \partial\mathcal{O} , \tag{4.6}$$

where $M(x)$ is a maximal positive subspace of R^N for $a_n(x)$, that is, a subspace maximal with respect to the property $a_n(x)\phi.\phi > 0$ for $0 \neq \phi \in M(x)$. There are infinitely many maximal positive subspace for $a_n(x)$. Which maximal positive subspace is the "right" one for A_o ?

The answer is simple. $M(x)$ should be chosen as the subspace spanned by the eigenvectors for positive eigenvalues of the symmetrizable matrix $b_n(x)^{-1} a_n(x)$, where

$$b_n(x) = \sum_{j,k=1}^{m} n_j(x)n_k(x)b_{jk}(x) > 0 , \quad x \in \partial O , \tag{4.7}$$

with $\{n_j(x)\}$ the unit outer normal at $x \in \partial\mathcal{O}$.

For the proof, let A_o be the m-accretive operator determined form A by this particular boundary condition. We first note that smooth functions satisfying this boundary condition form a core for A_o ; this is a result of a regularity theorem for A_o (see [6], see also the end of Example 8). Choosing such a function u , it then suffices to choose the boundary layer w_ε in the following way. After a preliminary localization and coordinate transformation, we may assume that the boundary is a part of the plane $x_m = 0$, with the domain $\mathcal{O}$ on the side $x_m > 0$, so that $a_n(x) = - a_m(x)$ and $b_n(x) = b_{mm}(x)$. Then set

$$w_\varepsilon(x) = \exp \left[\varepsilon^{-1}x_m b_{mm}(x',0)^{-1}a_m(x',0)\right] u(x',0) , \tag{4.8}$$

where $x' = (x_1,\ldots,x_{m-1}) \in R^{m-1}$. Note that the exponential matrix in (4.8) is small for $x_m > 0$ because $u(x',0) \in M(x)$ and $M(x)$ is a negative subspace for $a_m(x) = - a_n(x)$.

It is now easy to verify that (4.2) holds with S' the same formal differential operator as S without boundary condition. The required cancellation occurs due to the choice of w_ε as in (4.8). This proves the required resolvent convergence (1.1). Of course this implies the semigroup convergence (1.4) too. A similar result was given by Bardos, Brezis, and Brezis [13], where weak resolvent convergence is proved for more general differential operators.

References

[1] T. Kato - Lecture Series in Differential Equations. Vol.II, Van Nostrand Math. Studies, 1969, p.115-124.

[2] N. Okazawa - J. Math. Soc. Japan 27 (1975), p.160-165.

[3] L. Cattabriga - Rend. Sem. Mat. Univ. Padova 31 (1961), p.1-33.

[4] O. Ladyzhenskaya - The mathematical theory of viscous incompressible flow, 1961.

[5] T. Kato - J. Fac. Sci. Univ. Tokyo, Sec. I, Vol.17, 1970, p.241-258.

[6] D.S. Tartakoff - Indiana Univ. Math. J. 21 (1972), p.1113-1129.

[7] C.Y. Lai - Thesis, Berkeley 1975.

[8] N. Levinson - Ann. Math. 51 (1950), p.428-445.

[9] O. Ladyzhenskaya - Vestnik Leningrad. Univ. No.7 (1957), p.104-120.

[10] M.I. Vishik and L.A. Lyusternik - Uspehi Mat. Nauk 12 (1957), p.3-122.

[11] J.L. Lions - Contributions to Nonlinear Functional Analysis, Academic Press, 1971, p.523-564.

[12] P.D. Lax and R.S. Phillips - Comm. Pure Appl. Math.13 (1960), p.427-455.

[13] C. Bardos, D. Brezis and H. Brezis - Arch. Rational Mech. Anal. 53 (1973), p.69-100.

LES EQUATIONS SPECTRALES EN TURBULENCE HOMOGENE ET ISOTROPE. QUELQUES RESULTATS THEORIQUES ET NUMERIQUES.

M. LESIEUR et P.L. SULEM
Centre National de la Recherche Scientifique
Observatoire de Nice
B.P. 252, 06007 - Nice

RESUME

On considère des solutions aléatoires des équations de Navier-Stokes à trois dimensions qui sont statistiquement invariantes par translation d'espace (turbulence homogène), rotation et symétrie plane ; on obtient, moyennant certaines approximations, des équations intégro-différentielles non linéaires pour le spectre d'énergie (relié à la transformée de Fourier spatiale de la covariance des vitesses). Sur ces équations "spectrales", on peut mettre en évidence la régularité globale en temps pour toute viscosité positive et, dans un cas particulier, l'existence d'une singularité au bout d'un temps fini à viscosité nulle. Des résultats numériques à très faible viscosité sont présentés.

Le problème de la turbulence homogène se présente comme l'étude des solutions aléatoires des équations de Navier-Stokes (N.S.) considérées dans tout l'espace R^3 ,

$$\frac{\partial u}{\partial t} + (u.\nabla)u = -\nabla p + \nu\nabla u$$

$$\nabla.u = 0 \qquad (1)$$

$$u(\vec{x},0) = u_0\ (\vec{x}) \qquad \text{(aléatoire)}.$$

On peut rajouter éventuellement un terme de force extérieure également aléatoire. On se contente souvent d'une description partielle des solutions à l'aide de moments c'est-à-dire de moyennes (notées $<\cdot>$) de produits de vitesses en plusieurs points. Le moment d'ordre un $<u>$ est en général pris nul en turbulence homogène. Le tenseur des moments simultanés du second ordre $U_{ij}(\vec{x}_1,\vec{x}_2,t) = <u_i(\vec{x}_1,t)u_j(\vec{x}_2,t)>$, appelé covariance, est facilement mesurable et donne des renseignements sur l'énergétique de la turbulence. Dans le cas d'une turbulence homogène, isotrope et sans hélicité (c'est-à-dire invariante par translation d'espace, rotation et symétrie plane), ce tenseur est complètement caractérisé par sa trace $\sum_i U_{ii}(|\vec{r}|,t) = \sum_i <u_i(\vec{x}+\vec{r},t)u_i(\vec{x},t)>$ (Batchelor 1953). Notant $\hat{U}(k,t)$ la transformée de Fourier spatiale de cette dernière quantité (k = nombre d'onde = module de la variable de Fourier), on vérifie que

$$\text{l'énergie} = \frac{1}{2} <u^2(\vec{r},t)> = \frac{1}{2}\int_{R^3} \hat{U}(k,t)\ d^3\vec{k} \ .$$

On définit alors (à trois dimensions) le <u>spectre d'énergie</u> par

$$E(k,t) = 2\pi k^2\ \hat{U}(k,t) \quad ;$$

il vient

$$\frac{1}{2} <u^2> = \int_0^\infty E(k,t)\ dk \quad .$$

Le spectre d'énergie (qui, d'après le théorème de Bochner, est une quantité positive) décrit en quelque sorte la répartition de l'énergie entre les diverses échelles du mouvement.

Si l'on cherche à tirer des équations de Navier-Stokes des équations satisfaites par les moments de la vitesse, on se heurte à un "<u>problème de fermeture</u>" : de (1) on peut tirer des équations reliant les moments d'ordre n à ceux d'ordre $n + 1$, mais sans possibilité de se ramener à un système fini. Une façon d'éluder ce problème est alors d'imposer arbitrairement une "<u>hypothèse de fermeture</u>" entre certains moments. La plus connue est l' "Approximation Quasi-Normale" (Q.N.) introduite par Millionschtchikov (1941), où l'on suppose que les moments d'ordre quatre s'expriment en fonction des moments d'ordre deux comme si u était gaussien.

Cette approximation, qui a été étudiée analytiquement par Proudman et Reid (1954), a le grave défaut de ne pas préserver la positivité du spectre d'énergie (Ogura 1963). En fait, l'approximation Q.N. peut être corrigée de ce défaut en tenant compte, par une analyse phénoménologique simple, de l'intéraction avec les moments d'ordre plus élevé (Orszag et Kruskal 1968 ; Orszag 1974 ; Sulem, Lesieur et Frisch 1975). A trois dimensions, l'équation pour le spectre d'énergie s'écrit alors

$$\frac{\partial}{\partial t} E(k,t) + 2\nu k^2 E(k,t) = \iint_{\Delta_k} \theta_{kpq}(t) \frac{k}{pq} b_{kpq} [k^2 E(p,t)E(q,t) - p^2 E(q,t)E(k,t)] dp\, dq$$

$$E(k,t) \geqslant 0 \quad ; \quad k \geqslant 0 \qquad (2)$$

$E(k,0) = E_0(k)$ donné (en général à décroissance rapide pour $k \to \infty$).

L'intégrale porte sur le domaine Δ_k du plan p,q tel que k,p,q puissent former les côtés d'un triangle. Le coefficient b_{kpq} est donné par

$$b_{kpq} = \frac{p}{k} (xy + z^3) \quad ,$$

où x,y,z sont les cosinus des angles intérieurs du triangle k,p,q . La quantité $\theta_{kpq}(t)$ appelée "temps de relaxation des corrélations triples" est positive et complètement symétrique en k,p,q ; divers choix sont possibles : - le choix le plus simple $\theta_{kpq}(t) = \theta_0$ = constante correspond au MRCM (Markovian Random Coupling Model ; Frisch, Lesieur et Brissaud 1974). Il n'est physiquement pas très réaliste mais permet, comme on le verra, de pousser assez loin l'analyse mathématique ;
- une analyse phénoménologique plus conforme à la réalité conduit à prendre

$$\theta_{kpq}(t) = \frac{1 - \exp\{-[\mu(k,t) + \mu(p,t) + \mu(q,t)]\, t\}}{\mu(k,t) + \mu(p,t) + \mu(q,t)}$$

avec $\mu(k,t) = \nu k^2 + c^{te} \left(\int_0^k p^2 E(p,t)\, dp\right)^{1/2}$. C'est le EDQN (Eddy Damped Quasi Normal; Leith 1971 ; voir aussi Orszag 1974 ; Pouquet et al. 1975) ; - enfin, dans le TFM (Test Field Model ; Kraichnan 1971, Sulem et al. 1975) $\theta_{kpq}(t)$ apparaît comme solution d'une équation supplémentaire déduite d'une analyse beaucoup plus approfondie. Cette dernière approximation est en très bon accord avec les simulations numériques directes des équations de Navier-Stokes aléatoires (Orszag et Patterson 1972).

Dans tous les cas la préservation de la positivité du spectre $E(k,t)$ est assurée. Parmi les procédés qui permettent de le montrer, un des plus instructifs (mais non le plus simple) est fourni par la méthode des <u>modèles stochastiques</u> (Kraichnan 1961 ; Herring et Kraichnan 1972 ; Frisch, Lesieur et Brissaud 1974) : on démontre en effet que l'équation (2) peut s'obtenir comme conséquence <u>exacte</u> d'un modèle probabiliste obtenu en modifiant les termes non linéaires des équations de Navier-Stokes par des coefficients de couplage aléatoires. Dans le cas du MRCM, ce modèle s'écrit :

$$\frac{\partial u^\alpha}{\partial t} + \frac{1}{N} \sum_{\beta,\gamma=1}^{N} \Phi_{\alpha\beta\gamma}(t)\ (u^\beta\cdot\nabla)u^\gamma = -\nabla p^\alpha + \nu\Delta u^\alpha$$

$$\nabla\cdot u^\alpha = 0$$

où les indices α, β, γ varient de 1 à N . Pour α, β, γ fixés, les coefficients $\Phi_{\alpha\beta\gamma}(t)$ sont des bruits blancs gaussiens convenablement choisis, de valeur moyenne nulle et de covariance

$$<\Phi(t)\Phi(t')> = \theta_0\delta(t-t')$$

On montre alors que la quantité

$$E(k,t) = \lim_{N\to\infty} \frac{1}{N} \sum_{\alpha=1}^{N} E^\alpha(k,t)$$

où $E^\alpha(k,t)$ est le spectre d'énergie associé à la "réalisation" u^α , satisfait l'équation (2) . Dans le cas EDQN, les coefficients de couplage Φ dépendent en outre des nombres d'onde k, p, q .

L'étude de l'équation integrodifférentielle (2) nécessite l'introduction des quantités

$$|E(t)|_s = \int_0^\infty k^{2s}\ E(k,t)\ dk$$

qui jouent ici un rôle analogue à celui des normes de Sobolev. $|E(t)|_0 = \frac{1}{2} <u^2>$ est l'énergie et $|E(t)|_1 = \frac{1}{2} <(\mathrm{Rot}\ u)^2>$ est appelé "enstrophie" à cause du rôle particulier qu'elle joue en dimension deux où elle est conservée. On établit alors les estimations à priori suivantes pour des solutions suffisamment régulières (c'est-à-dire décroissant assez vite pour $k \to \infty$).

- Equation d'énergie

$$\frac{d}{dt} |E(t)|_0 + 2\nu|E(t)|_1 = 0 \quad , \tag{3}$$

valable pour le MRCM , le EDQN et le TFM.

- Equation ou inéquation d'enstrophie

$$\frac{d}{dt} |E(t)|_1 + 2\nu|E(t)|_2 \leqslant c^{te}\ |E(t)|_1^{3/2} \tag{4}$$

dans le cas du EDQN (André et Lesieur 1975),

$$\frac{d}{dt} |E(t)|_1 + 2\nu|E(t)|_2 = \frac{2}{3}\theta_0\ |E(t)|_1^2 \tag{5}$$

dans le cas du MRCM (Lesieur 1973).

De ces estimations à priori et de l'inégalité

$$|E(t)|_2 \geq |E(t)|_1^2 / |E(t)|_0 ,$$

on déduit,pour des solutions telles que initialement tous les $|E(0)|_n$ soient finis, les résultats suivants :

a) dans le cas du EDQN et du MRCM ,

1- pour toute viscosité positive, l'enstrophie reste bornée sur $[0,\infty[$. Il en résulte la régularité globale (en temps) et l'unicité de la solution (Bardos, Penel, Frisch et Sulem 1975) ;

2- à viscosité nulle, l'enstrophie reste bornée dans tout intervalle $[0,t]$, $t < t_*$, et la solution reste régulière sur cet intervalle [à rapprocher des résultats d'Ebin et Marsden (1970) et Kato (1972) sur l'équation d'Euler tridimensionnelle].

b) dans le cas du MRCM, on a en plus la propriété suivante : à viscosité nulle l'enstrophie devient infinie au bout d'un temps fini. Ce résultat de singularité est une conséquence immédiate de (5) .

On peut ensuite poser le problème du comportement de la solution du problème visqueux lorque $\nu \to 0$. Sur $[0,t]$, $t < t_*$, celle-ci tend vers la solution du problème inviscide $(\nu = 0)$, mais pour $t > t_*$ nous n'avons obtenu pour l'instant aucun résultat mathématique rigoureux (les problèmes analogues qui se posent sur l'équation de Burgers MRCM sont cependant réglés: Brauner, Penel et Temam 1974 ; Penel 1975 ; Bardos et al. 1975) . Le problème a néanmoins pu être étudié numériquement (André et Lesieur 1975) .

La Fig. 1 montre, dans le cas EDQN, l'évolution temporelle de l'énergie $|E(t)|_0$ pour des valeurs de plus en plus faibles de la viscosité : il apparait que, lorsque $\nu \to 0$, l'énergie n'est conservée que pendant un temps fini t_* , après quoi une viscosité infinitésimale suffit à provoquer une dissipation finie de l'énergie ("catastrophe énergétique" : Brissaud et al. 1973 ; Foias et Penel 1975).

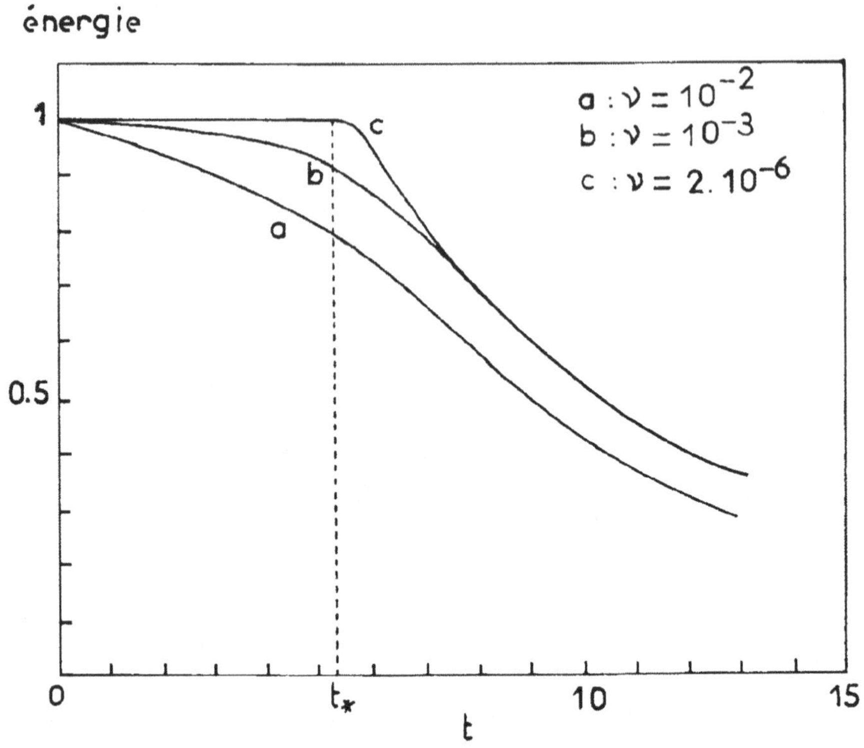

Fig. 1 Evolution temporelle de l'énergie $\frac{1}{2}$ $\langle u^2(t) \rangle$ pour des valeurs de plus en plus faibles de la viscosité (dans le cas EDQN). Noter que même aux très faibles viscosités, l'énergie est fortement dissipée après t_* .

La Fig. 2 montre, pour $\nu = 2.10^{-6}$, l'évolution temporelle d'un spectre d'énergie initialement à décroissance rapide avec une énergie et une enstrophie de l'ordre de l'unité. On constate un transfert de l'énergie vers des nombres d'onde de plus en plus grands au cours du temps. A l'instant t_* il apparaît une "zone inertielle" où $E(k,t)$ est proportionnel à $k^{-5/3}$, ce qui est conforme à la prédiction faite par Kolmogorov (1941) et Obukhov (1941) à partir de considérations phenoménologiques essentiellement dimensionnelles. Cette zone est limitée aux grands nombres d'onde par une "zone de dissipation" qui est rejetée de plus en plus loin quand la viscosité tend vers zéro. On remarque que le spectre en $k^{-5/3}$ s'étendant alors jusqu'à l'infini, l'enstrophie diverge.

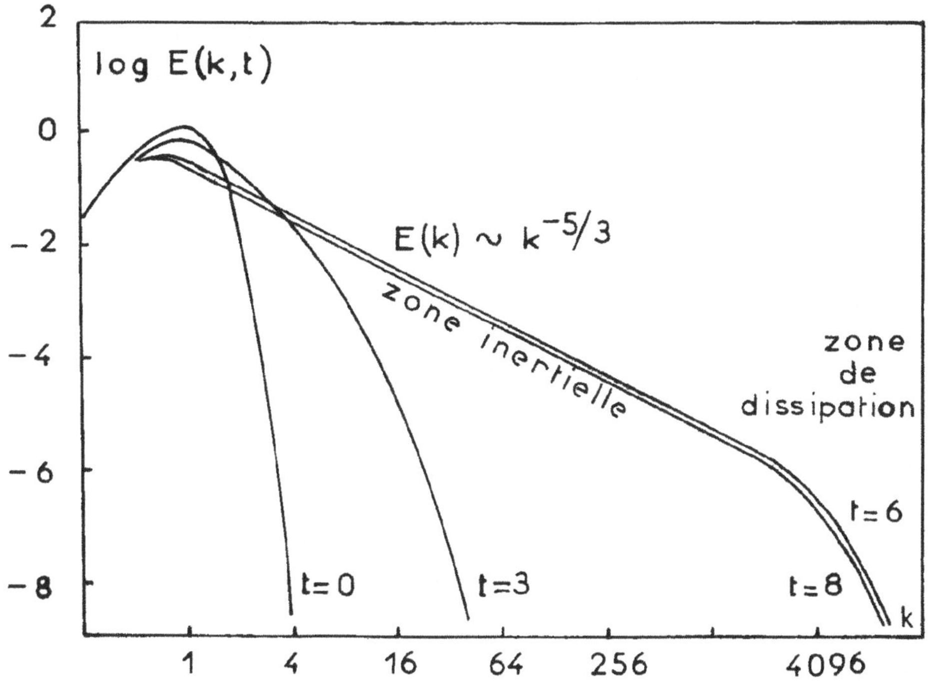

Fig. 2 Evolution temporelle du spectre d'énergie $E(k,t)$ dans le cas EDQN. [$\nu = 2.10^{-6}$, $E_0(k) \sim k^4 \exp(-2k^2)$] . Noter l'apparition après l'instant t_* , d'une zone "inertielle" en $k^{-5/3}$, et son évolution relativement lente (régime "quasi-stationnaire") .

Ces résultats numériques font donc apparaître pour $t \geqslant t_*$ et $\nu \to 0$ une limite singulière du spectre présentant une décroissance algébrique aux grands nombres d'onde.

Dans le cas de la turbulence bidimensionnelle, où l'on peut encore écrire une équation spectrale analogue à l'équation (2) , les phénomènes sont sensiblement différents du cas tridimensionnel : on a en plus de l'équation d'énergie ,

- l'équation d'enstrophie

$$\frac{d}{dt}\,|E(t)|_1 + 2\nu|E(t)|_2 = 0 \quad .$$

La conservation de l'enstrophie par les termes non linéaires exclut cette fois la possibilité d'une catastrophe énergétique. Quant à $|E(t)|_2$, on peut montrer qu'il croît au plus exponentiellement, ce qui assure la conservation de l'enstrophie dans la limite de viscosité nulle (Pouquet, Lesieur, André et Basdevant 1975). Ce résultat est à rapprocher de ceux obtenus sur la régularité de l'équation d'Euler bi-dimensionnelle (Ebin et Marsden 1970).

BIBLIOGRAPHIE

ANDRE, J.C. et LESIEUR, M. 1975 : "Evolution of High Reynolds Number Isotropic Helical Turbulence". Prépublication. Observatoire de Nice.

BARDOS, C., PENEL, P., FRISCH, U. et SULEM, P.L. 1975 : "Modified Dissipativity for a non-linear Evolution Equation Arising in Turbulence". Compte-rendu de ce Colloque.

BRAUNER, C.M., PENEL, P. et TEMAM, R. 1974. C.R. Acad. Sci. Paris, A279, p 65 et 115.

BATCHELOR, G.K. 1953. "The Theory of Homogeneous Turbulence". Cambridge University Press.

BRISSAUD, A., FRISCH, U., LEORAT, J., LESIEUR, M., MAZURE, A., POUQUET, A., SADOURNY, R. et SULEM, P.L. 1973. Ann. Géophys., 29 , 539.

EBIN et MARSDEN, 1970. Annals of Mathematics 92 , 102.

FOIAS, C. et PENEL, P., 1975 . C.R. Acad. Sci. Paris A280 , 629 .

FRISCH, U., LESIEUR, M. et BRISSAUD, A. 1974. J. Fluid. Mech. 65 , 145 .

HERRING, J.R. et KRAICHNAN, R.H. 1972. Dans "Statistical Models and Turbulence", ed. M. Rosenblatt et C. Van Atta.,Springer, p. 148

KOLMOGOROV, N.A. 1941. C. R. acad. Sci. URSS. 30 , 301.

KATO, T., 1972 . J. Funct. Anal. 9 , 296.

KRAICHNAN, R.H. 1961, J. Math. Phys. 2 , 124. erratum 3 , 205 (1962).

KRAICHNAN, R.H. 1971, J. Fluid. Mech. 47 , 513.

LESIEUR, M. 1973, Thèse, Université de Nice.

LEITH, C.E. 1971. J. Atmos. Sci. 28 , 145.

MILLIONSCHTCHIKOV, M. 1941. C.R. Acad. Sci. URSS 32 , 615.

OBUKHOV, A.N. 1941, C.R. Acad. Sci. URSS, 32, 19.

OGURA, Y. 1963; J. Fluid. Mech. 16 , 33.

ORSZAG, S.A. 1974. "Lecture notes on the statistical theory of turbulence" . Cours à l'Ecole d'Eté de Physique Théorique des Houches 1973. Gordon and Breach.

ORSZAG, S.A. et KRUSKAL, M.D. 1968. Phys. Fluids 11 , 43 .

ORSZAG, S.A. et PATTERSON, G.S. 1972 : dans "Statistical Models and Turbulence" ed. M. Rosenblatt et C. Van Atta, Springer, p.127.

PENEL, P. 1975. Thèse, Université de Paris Sud.

POUQUET, A., LESIEUR, M., ANDRE, J.C. et BASDEVANT, C. 1975 : "Evolution of High Reynolds Number Two-dimensional Turbulence". A paraître dans J. Fluid. Mech.

PROUDMAN, I. et REID, W.H. 1954. Phil. Trans. Roy. Soc. A247, 163.

SULEM, P.L., LESIEUR, M. et FRISCH, U. 1975 : "Le Test-Field-Model interprété comme méthode de fermeture des équations de la turbulence" A paraître dans Ann. Geophys.

INTERMITTENT TURBULENCE AND FRACTAL DIMENSION: KURTOSIS AND THE SPECTRAL EXPONENT 5/3+B

Benoit Mandelbrot

General Sciences Department
IBM Thomas J. Watson Research Center
POBox 218, Yorktown Heights, New York 10598, USA

Various distinct aspects of the geometry of turbulence can be studied with the help of a wide family of "shapes", for which I have recently coined the neologism "fractals". Until now, they had been used hardly at all in concrete applications, but I have shown them to be useful in a variety of fields. In particular, they play a central role in the study of a) homogeneous turbulence, through the shape of the iso-surfaces of scalars (Mandelbrot 1975a), b) dispersion (Mandelbrot 1976a), and especially c) the intermittency of dissipation (Mandelbrot 1972 and 1974a,b). Fractals are all loosely characterized as being violently convoluted and broken up, a feature denoted in Latin by the adjective "fractus". Fractal geometry approaches the loose notion of "form" in a manner different and almost wholly separate from the approach used by topology.

The present paper will sketch a number of links between the new concern with fractal geometry and the traditional concerns with various spectra of turbulence and the kurtosis of dissipation. Some of the results sketched will lead to improvement and/or correction of results found in the literature, including further refinement of Mandelbrot 1974a,b.

One result described in Chapter IV deserves special emphasis: It confirms that intermittency requires that the classical spectral exponent 5/3 be replaced by 5/3+B. However, the factor B turns out in general to be different from the value ordinarily accepted in the literature, for example in the treatise by Monin and Yaglom 1975. Said value, derived by Kolmogorov, Obukhov, and Yaglom, is linked to the so-called lognormal hypothesis, which is a separate *Ansatz*, and is highly questionable.

Other results in the paper are harder to state precisely in a few words. In rough terms, they underline the convenience and heuristic usefulness of the fractally homogeneous approximation to intermittency, originating in Berger & Mandelbrot 1963 and Novikov & Stewart 1964. On the other hand, they underline the awkwardness of the lognormal hypothesis. That it is only an approximation has been stressed by many writers of the Russian school, but it becomes increasingly clear that even they underestimated its propensity to generate paradoxes and to hide complexities.

My book *Les objets fractals: forme, hasard et dimension*, Mandelbrot 1975b, describes numerous other concrete applications of fractals. It can also serve as a general background reference, but its Chapter on turbulence is too skimpy to be of use here. This deficiency should soon be corrected in the English version, tentatively titled *Fractals: form, chance and dimension,* which is being specifically designed to also serve as preface to technical works such as the present one. Nevertheless, in its main points, the present text is self-contained.

I. CURDLING AND FRACTAL HOMOGENEITY. ROLE OF THE FRACTAL DIMENSION $D=2$.

The term "curdling" is proposed here to designate any of several cascades through which dissipation concentrates in a small portion of space. Absolute curdling is described by the Novikov & Stewart 1964 cascade. Its outcome was described independently, without any generating mechanism, in Berger & Mandelbrot 1963 and Mandelbrot 1965. Weighted curdling is described by the cascades of Yaglom 1966, Mandelbrot 1974b and Mandelbrot 1972. It turns out that absolute curdling is more realistic than suggested by its extreme simplicity and in addition it provides an intrinsic point of reference to all other models. Therefore it deserves continuing attention. Weighting will be examined next. It turns out that it mostly adds complications, and its apparent greater generality is in part illusory.

Absolute curdling. Before each stage, dissipation is assumed uniform over a certain number of spatial cells, and zero elsewhere. Curdling concentrates it further: each of the initial cells breaks into $C = \Gamma^3$ sub-cells and dissipation concentrates within $N \geq 2$ of these, called "curds". The quantity $\rho = 1/\Gamma$ is the ratio of similarity of sub-cells with respect to the cells.

After a finite number of stages of absolute curdling, dissipation concentrates with uniform density in a closed set, whose outer and inner scales are L and η and which constitutes an approximation to a fractal. Figure 1 represents such an approximate fractal in the plane. (We shall soon see it is nearly a plane cut through an approximate spatial fractal.)

The most important characteristic number associated with a fractal is its fractal (Hausdorff) dimension, which in the present case is most directly defined as

$$D = \log N/\log (1/\rho).$$

It is always positive (because of the condition $N \geq 2$), and it is ordinarily a fraction. On Figure 1, $\Gamma = 5 = 1/\rho$ and $N = 15$, so that $D = 1.6826$ (the values of N and ρ were chosen to make this D as close as conveniently feasible to $5/3$). One reason for calling D a dimension is elaborated upon in Figure 2.

The notion of fractal dimension also extends to shapes that are not self similar. D is never smaller than the topological dimension D_T. For the classical shapes in Euclid, $D = D_T$ and the shapes for which I had introduced the term fractals are by definition such that $D > D_T$.

When dissipation is uniform over such a fractal of dimension $D<3$, turbulence will be called *fractally homogeneous*. The modifier is of course meant to contrast it with G. I. Taylor's classical concept of homogeneous turbulence, which can henceforth be viewed as the special limit case of fractally homogeneous turbulence for $D \to 3$. The salient fact is that the generalization allows $D-3$ to be negative.

The value of D refers to just one among many mathematical structures. It follows that the same D can be encountered in sets that differ greatly from other viewpoints, for example are topologically distinct. Nevertheless, many aspects of fractally homogeneous turbulence turn out to depend solely upon D. In an approximate fractal of dimension D and scales L and η, it is clear that dissipation concentrates within $(L/\eta)^D$ out of $(L/\eta)^3$ cells of side η. This approximation's volume is $(L/\eta)^D \eta^3$. The relative occupancy ratio of the region of dissipation (measured by the relative number of curds of side η within a cell of side L) is $(\eta/L)^{D-3}$. Therefore the uniform density of dissipation in a curd must be equal to $(L/\eta)^{3-D}$ times the overall density of dissipation.

The difference $3-D$ (or, for sets in Δ-dimensional Euclidean space with $\Delta \neq 3$, the difference $\Delta - D$) will be called *codimension*. This usage is consistent with the usage prevailing in the theory of vector spaces.

The quantities evaluated in the preceding paragraph concern solely the way blobs of intermittent turbulence *spread around*. Therefore D is a so-called metric characteristic. It is conceptually distinct from the topological characteristics of the way blobs are *connected*. And indeed, as is already the case for defining inequality $D > D_T$, most of the relationships between fractal and topological structures are expressed by inequalities. Topological structures prove very difficult to investigate, but fractal structures are more easily accessible to analysis. It is fortunate therefore, as we shall show in this paper, that several structures which a casual examination would classify as topological actually turn out to be exclusively or predominantly metric, namely, fractal. One example is the degree of intermittency as measured by the kurtosis. Another is the intermittency correction to the $2/3$ and $5/3$ laws, even though intermittency may conceivably have a distinct topological facet.

Weighted curdling. This more general process proceeds as follows. Each stage starts with dissipation uniform within cells. Then the density of dissipation in each subcell of a cell is multiplied by a random factor W, with $\langle W \rangle = 1$. As a concept, W is related, but not identical, to the Yaglom multiplier. The cascade underlying weighted curdling is a generalization and a conceptual tightening-up of various arguments concerning the lognormal distribution. (In the absence of viscosity cutout, weighted curdling leads asymptotically to an everywhere dense fractal that is – topologically – open rather than closed. However, this distinctive feature vanishes when high frequencies are cut out by viscosity.)

The inner scale. The value of η, of course, is determined by the dissipation and the viscosity ν. Its value in the Taylor homogeneous case is well known. In the fractally homogeneous case, it continues to be well-defined, with a value that also depends on D. In the general case, however, the notion of η involves great complications. They will be avoided in Chapters III and IV, and only faced in Chapter V.

Behavior of linear cross-sections and a deep but elementary experimental reason to believe that for turbulent dissipation $D>2$. Most conveniently, the fractal dimensions of the linear and planar cross-sections of a fractal are given by the same formulas as the Euclidean dimension of the corresponding cross-sections of an elementary geometric shape. We shall state the rule, then show that, combined with evidence, it suggests very strongly that the D of turbulent diffusion must be greater than 2.

Rule: When the fractal or Euclidean dimension D of a shape is above 2, then its cross-section by an arbitrarily chosen straight line has a positive probability of being non-empty with the dimension $D-2$. Otherwise, the cross-section is empty. For planar sections analogous results apply, except that, instead of subtracting 2, one must subtract 1 (Mattila 1975). *Finite η-approximations to fractals.* Start with $(L/\eta)^D$ curds of side η. When $D>2$, the typical line cross-section will either be near-empty, or (with a positive probability, near independent of η) will include about $(L/\eta)^{D-2}$ segments of side about η. When $D<2$, on the contrary, the probability of hitting more than a small number of curds, say two curds or more, will greatly depend on η and will tend to zero with η/L. At the limit, suppose that $D=\log 2/\log(L/\eta)$ (which, among possible values of D, is the closest to $D=0$); then everything concentrates in two curds; the probability of hitting either by an arbitrarily selected line or a plane is minute. *Illustration.* Figure 1, being of dimension $\log 15/\log 5$, has the same dimension as the typical planar section of an approximate spatial fractal with $D=1+\log 15/\log 5$.

Application. By necessity, turbulence is ordinarily studied through linear cross-sections in space-time. Under Taylor's frozen turbulence assumption, they are the same as linear cross-sections through space. Turbulence is a highly prevalent phenomenon, in the sense that the typical cross-section hits it with no effort and repeatedly. Such would not be the case if the cross-sections were almost surely empty. Hence, we have an elementary reason (and hence an especially profound one) for believing that the fractal dimension D of turbulent dissipation satisfies $D>2$.

Digression. Possible relevance of fractal geometry to the study of the Navier-Stokes and Euler equations. My approach to the geometry of turbulence is to a large extent "phenomenological", as was Kolmogorov's approach, and is geometric rather than dynamic. It cannot rely on any information drawn from the study of the Navier-Stokes and Euler equations. However, the converse could be true: any success the fractal approach may be able to achieve should assist in the notoriously difficult search for turbulent solutions. I think, indeed, that the greatest roadblock in this search has been due to the lack of an intrinsic characterization of what was being sought. One could even go as far as to argue that no one could be sure he would recognize such a solution if it were shown to him. In past studies of other equations of physics, on the contrary, the easiest procedure has frequently been to seek guidance in guesses concerning the singularities to be expected in the solution. Knowing what to look for has often made it less difficult to find it, but this approach has not yet worked for turbulence. Von Neumann 1949-1963 has noted that "its mathematical peculiarities are best described as new types of mathematical singularities", but he made no progress in identifying them.

In this vein, I propose to infer from empirical evidence that, for nonlinear partial differential equations like Euler's system (when viscosity is absent) or the Navier-Stokes system (when viscosity $\rightarrow 0$, or possibly even at a positive small viscosity), the singularities of sufficiently "mature" solutions are likely to tend towards being fractal.

The singularities of Euler solutions should be viewed as associated with curdling, as discussed above and in the body of this paper. As to the Navier-Stokes equations, the notion that the solution can possess singularities remains unproven and in fact controversial, but if singularities in the Oseen-Leray sense do in fact exist, they must be enormously "sparser" than the Eulerian ones; possibly a proper subset. Assuming they indeed exist, Scheffer 1975 has been successful in restating some of my rough hunches on this topic into precise conjectures, and has proved several of them, relating them with the work of J. Leray 1934 and opening new vistas on this ancient problem. See also Scheffer's paper in the present volume.

Given that closely related forms of intermittency are found to occur in phenomena ruled by diverse other equations, the specific characteristic of the Navier-Stokes equation, which leads to the "fractality" of the solutions, must have its counterpart in broad classes of other equations, and it may well be more useful to study them within a broader mathematical context.

Digression. A second connection between fractality and Navier-Stokes equations. This connection comes in through the shapes of coastlines. A priori, it may well be that fractality is *wholly* related to the Oseen-Leray argument that a solution with good initial data may, after a certain time, have large velocity gradients. Alternatively, we have the Batchelor & Townsend (1949) argument "that the distribution of vorticity is made 'spotty' in the early stages of the decay by some intrinsic instability and is kept 'spotty' throughout the decay by the action of the quadratic terms of the Navier-Stokes equations". However, "spottiness" may also be affected by a third factor. Indeed, the study of partial differential equations, while stressing the respective roles of the equation itself and of boundary conditions, usually fails to consider the possible effects of the shape of that boundary. More precisely, the boundary is nearly always assumed very smooth, for example is taken to be a cube. For atmospheric and ocean turbulence, this approximation may well be unrealistic. The fact is as I showed in Mandelbrot 1967 and in the book *Fractals*) that the shapes of coastlines contain features whose "typical lengths" cover a wide span, so that their fractal dimensions are greater than 1. This and the analogous statement concerning the rough surface of the Earth may well combine with intrinsic instabilities as a third contribution to the roughness of observed flow.

II. THE FUNCTION $f(h)$. THE FRACTAL DIMENSIONS ARE DETERMINED BY $f'(1)$.

All the aspects of intermittency to be studied in this paper are ruled by power laws. If one adds specific further assumptions, the various exponents are linked to each other (through the fractal dimension D of the carrier or the parameter "μ" of the Kolmogorov theory). However, in the general case they are distinct. As Novikov 1969 had observed in the case of spectra and moments, each power law is merely a symptom of self similarity. The multiplicity of different exponents shows the self similarity syndrome to be complex and multifarious.

Nevertheless, the exponents that enter in my previous papers and in the present one can all be derived from various distinct properties of the following determining function:

$$f(h) = \log_C \langle W^h \rangle;$$

recall that $C = \Gamma^3$ is the number of subcells per cell. In absolute curdling, W is a binomial random variable: it can either vanish or take one other possible value $1/p$ with the probability p, so that $\langle W \rangle = 1$. In weighted curdling, W is a more general random variable, still satisfying $\langle W \rangle = 1$. Further, the limit lognormal model of Mandelbrot 1972 also fits in the same scheme by appropriate interpretation of W. By a general theorem of probability (Feller 1971, p.155), $f(h)$ is a convex function; it obviously satisfies $f(1)=0$. Furthermore, whenever $\Pr(W>0)=1$, one also has $f(0)=0$. As a first example, the graph of $f(h)$ is a straight line if, and only if, curdling is absolute. If so, the graph passes through $f(1)=0$ but not through $f(0)=0$. As a second example, $f(h)$ is a parabola when W is lognormal. These two cases are drawn on Figure 3, which also illustrates other features of f(h).

My past and present papers show that the following characteristics of f(h) are of interest: f(2/3), f'(1), f''(1), f(2), f(h) for h integer >2, and α_1, α_2, $\alpha_3 = \alpha$, where α_m is defined as the root other than h=1 of the equation $\phi_m(h) = 3f(h) - m(h-1) = 0$. The functions ϕ_m being convex, each α_m is unique, but of course one or more among them can be infinite. Since the condition $\phi_1(h)<0$ is at least as demanding as $\phi_2(h)<0$, and, a fortiori, as $\phi_3(h)<0$, we see that, if $\alpha_1>1$, one has $\alpha_1 \leq \alpha_2 \leq \alpha_3 = \alpha$.

The order in which these various characteristics have been listed in the preceding paragraph is that of increasing sensitivity to the detail of the distribution of W. The first and least sensitive – and, in my opinion, the most basic – is

$$f'(1) = \langle W \log_C W \rangle$$

This quantity was first considered in Mandelbrot 1974a,b, further results being due to Kahane 1974 and Kahane & Peyrière 1976. When $3f'(1)<3$, the carrier of intermittency is nondegenerate, and its fractal dimension is $D=3-3f'(1)$. Thus, $3f'(1) = \langle W \log_\Gamma W \rangle$ determines the codimension 3–D. As to the planar and linear intersections, when $3f'(1)<2$, respectively when $f'(1)<1$, these intersections are non-degenerate, with fractal dimensions equal to $D_2 = 2-3f'(1)$, respectively to $D_1 = 1-3f'(1)$. As expected, $D_2 = D-1$ and $D_1=D-2$. In the case of a lognormal W, and denoting by μ the basic parameter, one has $f(h) = (h-1)h\mu/6$. Hence, $D = 3-\mu/2$ and μ is merely twice the codimension.

Next, as the present paper will show, some other characteristics of f(h), more sensitive to details of W, rule the traditional concerns with second order (spectral) properties. One must distinguish an inertial and a dissipative range (these are probably misnomers). In the former, the value of f(2/3) rules the corrective term B to be added to the exponent in the Kolmogorov $k^{5/3}$ law (this will be shown in Chapter IV). Similarly, the value of f(2) rules the variance, the kurtosis and the exponent of the spectrum of dissipation (this will be shown in Chapter III). The next simplest characteristic of f(h) is f''(1). Our last result will be that f''(1) determines the width of the dissipative range. When f''(1)=0, an equality characteristic of the fractally homogeneous case, the dissipative range is vanishingly narrow. Otherwise, it is most significant, especially when L much exceeds the Kolmogorov inner scale.

Digression. Each property that involves a moment of higher order h is ruled by the corresponding f(h). As h increases, f(h) becomes increasingly sensitive to details of the distribution of W, which is why the moments computed from the lognormal assumption appear inconsistent; see Novikov 1969. This difficulty was eliminated, however, when Mandelbrot 1972,1974a,b showed that – in the absence of viscosity cutoff – the population moments above a certain order, namely α_1, α_2, or α, are in fact infinite. In my opinion, this feature explains why the experimentalists have found empirical moments of higher order to be so elusive.

III. COVARIANCE, FLATNESS AND KURTOSIS OF DISSIPATION. EXPONENTS DETERMINED BY f(2).

The exponent in the covariance of the dissipation. Take two domains Ω' and Ω'' whose diameters are small compared to the smallest distance r between them, and large compared to η. We define the covariance of the dissipation density $\varepsilon(\mathbf{x})$ as the expectation of the product of the average of $\varepsilon(\mathbf{x})$ within these domains. Without entering into details, let it be stated that in the fractally homogeneous case this covariance is approximately $(r/L)^{D-3}$, and in the general case it is $(r/L)^{-3f(2)}$. The proof follows closely that of Yaglom (see Monin & Yaglom 1975, p. 614), but stops before the point where these authors approximate the product of many W's by a lognormal variable.

By the convexity of f(h), $f(2) \geq f(1) + (2-1)f'(1) = f'(1)$. Thus, $3f(2) \geq 3f'(1) = 3-D$. Equality prevails if and only if f(h) is rectilinear, i.e., the curdling is absolute and turbulence is fractally homogeneous. (This is, in addition, the sole case where one can make r as low as η.) In every other case, an evaluation of the codimension 3–D through the observed exponent f(2) would lead to overestimation. For example, for the strictly lognormal W, $3f(2)=\mu$, which is the *double* of the estimate using the dimension, namely $3f'(1) = \mu/2$.

Let us now show that the same exponents play an equally central role in the study of the kurtosis of dissipation after averaging over small domains of side r. In other words, at least in intermittency generated by curdling, the covariance and kurtosis of dissipation are conceptually identical.

Kurtosis in the fractally homogeneous case. Here, we know that the dissipation vanishes, except in a region of relative size $(L/r)^{D-3}$, in which it equals $(L/r)^{3-D}$. Hence, it is readily shown that the kurtosis is simply $(L/r)^{3-D}$. It increases as r becomes smaller, and when r takes its minimum value η (as announced, we shall show that η is well-defined in the fractally homogeneous case), the kurtosis reaches its maximum value $(L/\eta)^{3-D}$. The measure of degree of intermittency depends both on the intrinsic characteristic of the fluid, as expressed by D, and on outer and inner scale constraints, as expressed by L/η, which is related to Reynolds number. Therefore, it is better to measure the degree of intermittency by D itself. The empirical value of the exponent is 0.4 (Kuo and Corrsin 1972), suggests under the assumption of fractal homogeneity that D=2.6.

To explore the significance of these findings, let me begin by sketching the results of previous studies of the kurtosis by Corrsin 1962 and by

Tennekes 1967. These and other authors took it for granted that the exponent of kurtosis depends mainly upon what may be called the proto-topological shape of the carrier of intermittency, namely on whether it is a "blob", a "slab", or a "sheet". While other assumptions also entered each model, they were felt to be secondary. This impression turns out to have been unwarranted. The crucial fact is that each of these models leads to fractally homogeneous intermittency, whose dimension D is affected by *all* the assumptions made, and determines the exponent of the kurtosis.

It turns out that, in the Corrsin model, the exponent's value is $3-D=1$ (his formula 10), hence $D=2$. This fractal dimension is experimentally wrong, in fact fails to satisfy the basic requirement $D>2$. It is interesting to note that $D=2$ is the smallest fractal dimension compatible with Corrsin's featured assumption, that turbulent dissipation concentrates with uniform density within sheets of thickness η enclosing eddies of size L. In other words, Corrsin's additional assumptions cancel out: there was no surreptitious increase of D, and he worked with a classical shape rather than with a fractal.

On the other hand, the exponent in the Tennekes model can be seen to imply $D = 7/3$. This value *does* satisfy $D>2$, and is reasonably close to observations. On the other hand, it very much exceeds the minimum fractal dimension, namely $D=1$, which topology imposes on a shape including ropes. Hence, Tennekes was mistaken in featuring the assumption that dissipation occurs in vortex tubes of diameter η. The more vital assumption was that the average distance between tubes is the Taylor microscale λ. The fact that $D = 7/3$ is even higher than the Corrsin value $D = 2$ strongly underlines that a tube, if sufficiently convoluted, ends up by ceasing to be a tube from a metric-fractal viewpoint, and becomes a fractal. Finally, the experimental $D = 2.6$ satisfies $D>2$. In addition, it *does not exclude* the presence of either ropes or sheets, but *does not require* either.

Kurtosis of nonfractally homogeneous intermittency generated by weighted curdling. In this case, kurtosis is simply $\langle\varepsilon^2\rangle/\langle\varepsilon\rangle^2 = \langle\varepsilon^2\rangle$ and turns out to be equal to

$$\langle W^2\rangle^{\log_\Gamma(L/r)} = (L/r)^{\log_\Gamma\langle W^2\rangle} = (L/r)^{3f(2)}$$

We know that $3f(2) \geq 3-D$. Hence, among all forms of turbulence generated by curdling and having a given D, the fractally homogeneous case is the one where the kurtosis is smallest. Hence $3f(2) = 0.6$ only yields $D \geq 2.6$.

Digression concerning the fractally homogeneous case. The behavior of the Fourier transform. Fourier transforms do not deserve the near exclusive attention which the study of turbulence gave to them at one time (through spectra), but they are important. It may be useful, therefore, to mention that, in the fractally homogeneous case, their properties happen to involve fractal dimension. The strength of the relationship has long been central to the fine mathematical aspects of trigonometric series see Kahane & Salem 1963, but the resulting theory has thus far been little known and used beyond its original context.

It deals particularly with functions that are constant except over a fractal of dimension D. Such functions have no ordinary derivative but have generalized derivatives which are measures carried by the fractal in question. The rough result is that the Fourier coefficients of the measure in question turn out in many typical cases to decrease like k^{-D}.

In a finer approximation, however, the considerations of fractal dimensionality are logically distinct from spectra. This fact further elaborates the assertion made earlier, that the consequences of the self similarity of turbulence split into conceptually distinct aspects, dimensional, spectral and others, which are governed by different exponents of self similarity linked to each other through inequalities.

If, as I hope, the importance of fractal shapes in turbulence becomes recognized, the spectral analysis of the motion of fluids may become able at last to make some use of a considerable number of pure mathematical results relative to harmonic analysis.

IV. THE MODIFICATION IN THE 2/3 AND 5/3 LAWS. A SPECTRAL EXPONENT CHANGE DETERMINED BY $f(2/3)$.

It has been noted in Kolmogorov 1962 and Obukhov 1962 that intermittency changes the exponents $5/3$ and $2/3$ by adding a positive factor to be denoted by B. A more careful examination of the problem, to which we now proceed, confirms this conclusion but yields a value of B that does not in general fit those asserted by the Russian school.

Consider two points P' and P'' separated by the distance r and denote the velocity difference $\mathbf{u}(P'') - \mathbf{u}(P')$ by $\Delta\mathbf{u}$. In Taylor homogeneous turbulence of constant dissipation denotes $\delta(\mathbf{x}) \equiv \varepsilon$, one has $\langle(\Delta\mathbf{u})^2\rangle = (\varepsilon r)^{2/3}$. To extend this result to the intermittent case, when the nonrandom ε is replaced by a random field $\varepsilon(\mathbf{x})$ with $\langle\varepsilon(\mathbf{x})\rangle = \varepsilon$, one must replace ε in $(\varepsilon r)^{2/3}$ by some quantity characteristic of said random field and also of P' and of P''. Even though this quantity may be determined in several different ways, there will be no harm in always denoting it by ε_r.

Like Yaglom, we shall first follow closely the approach of Obukhov 1962 and Kolmogorov 1962, who propose one should take as ε_r the average of $\varepsilon(\mathbf{x})$ over a sphere – we shall call it the Obukhov sphere – whose poles are P' and P''. We shall designate this domain as $\Omega(P', P'')$. In practice, in the case of curdling within cubic cells, Ω is more conveniently the smallest cell containing both P' and P. We shall find, as have Kolmogorov and Obukhov, that intermittency requires the replacement of the classical spectral density $E(k) = E_0\varepsilon^{2/3}k^{-5/3}$ by $E(k) = E_0 \varepsilon^{2/3}k^{-5/3}(k/L)^{-B}$. On the other hand, we shall disagree with them on two basic points: a), the value of the exponent B, and b), the highest value of k for which a spectral density with the exponent $5/3+B$ is conceivable.

When expressed in terms of the dimension D, the Kolmogorov-Obukhov correction comes out as $B = (3-D)/4.5$, but this value turns out to be due to very specific and arguable features of the lognormal assumption, which is part of their model. In the case of fractally homogeneous intermittency, one finds the different and *larger* value $B = (3-D)/3$, and in general $B = -3f(2/3)$, which can lie anywhere between the bounds 0 and $(3-D)/3$. Thus the value $(3-D)/4.5$ is a kind of compromise, perfectly admissible but by no means necessary.

Secondly, the fractally homogeneous case is unique in that it allows a widely-liked approximation in which the dissipative range reduces to a point, and one can assume that E(k) takes the above form all the way to the inverse of the proper inner scale, and vanishes beyond. In all other cases, this traditional approximation leads to paradox. This Chapter will evaluate B and present the paradox; Chapter V will resolve it.

The following Sections will derive the above result, then subject the approach of Obukhov-Kolmogorov to a critical analysis. Their choice for ε_r was indeed acknowledged to be to a large extent an arbitrary first trial suggested less by physics than by commodity. Other definitions of ε_r are therefore worth considering. The first alternative ε_r will be the average of $\varepsilon(\mathbf{x})$ over a domain Ω that is the interval P' P". When $D>2$, as we believe is the case for turbulence, the expression for B is unchanged and the coefficient E_0, while modified, remains positive and finite. When $D<2$, on the contrary, E_0 vanishes and B becomes meaningless. To obtain the second and last alternative ε_r, we shall let Ω be determined by the distribution of intermittency, and we shall thereby bring in topology. The argument will give reasons for believing that the carrier of turbulence, not only must satisfy the metric inequality $D>2$ proven in Chapter I, but must, in some topologic sense, be "at least surface-like". However, this third choice of Ω is very tentative, and so are the conclusions drawn from it.

1. THE FRACTALLY HOMOGENEOUS CASE WHEN Ω IS THE OBUKHOV SPHERE OR AN APPROXIMATING CUBE

In this case, ε_r is the average of $\varepsilon(\mathbf{x})$ over Ω. By the theorem of conditional probabilities, one can factor $\langle\varepsilon_r^{2/3}\rangle$ as the product of a) the probability of hitting dissipation in Ω, and, b) the conditional expectation of $\varepsilon_r^{2/3}$ where "conditional" means that averaging is restricted to the cases where $\varepsilon_r>0$.

When Ω is the Obukhov sphere or the smallest cubic eddy that contains both P' and P", it can be shown that, as $n\to\infty$, the hitting probability becomes at least approximately equal to $p_0(r/L)^{3-D}$.

Since the product of the hitting probability by the conditional expectation of ε_r is simply the nonconditional expectation ε, the conditional expectation must be equal to $\varepsilon\ (r/L)^{D-3}$. A stronger statement, in fact, holds true. Assuming fractal homogeneity, ε_r, when positive, it can be shown to be the product of $\varepsilon r^{D-3} L^{3-D}$ by a random variable having positive and finite moments of every order. Consequently,

$$\begin{aligned}\langle(\varepsilon_r r)^{2/3}\rangle &= V_{1/3}\,(r/L)^{3-D}\,\varepsilon^{2/3}\,r^{2/3}\,(r/L)^{-(3-D)/3}\\ &= V_{1/3}\,\varepsilon^{2/3}\,L^{-(3-D)/3}\,r^{2/3+(3-D)/3}\\ &= V_{1/3}\,\varepsilon^{2/3}\,L^{-(3-D)/3}\,r^{1-(D-2)/3}.\end{aligned}$$

The corresponding spectral density is

$$\begin{aligned}E(k) &= E_0\,\varepsilon^{2/3}\,L^{-(3-D)/3}\,k^{-5/3+(D-3)/3}\\ &= E_0\,\varepsilon^{2/3}\,L^{-(3-D)/3}\,k^{-2+(D-2)/3}.\end{aligned}$$

It is important to know that the numerical coefficients E_0 and $V_{1/3}$ are positive, but their actual values will not be needed.

These expressions show that intermittency has two distinct effects: to inject L and to change the exponent of k from $5/3$ to $5/3+B$, where $B = (3-D)/3$.

Since $B\geq 0$, the exponent $5/3+B$ always exceeds the 1941 Kolmogorov value $5/3$. As expected, $B=0$ corresponds to the limit case $D=3$, when dissipation is distributed uniformly over space.

The point where $5/3+B$ goes through the value 2 occurs when $D=2$, a relationship to which we shall return in Section 4.

Even if it is confirmed that (as inferred in Chapter I) ordinary turbulence satisfies $D>2$, it is good to include the values $D<2$ for the sake of completeness. Since B satisfies $B\leq 1$, the exponent $5/3+B$ always lies below the value $8/3$. This value is seen to correspond to $D=0$, the limit case when dissipation concentrates in a small number of blobs. (In absolute curdling, we saw that D is at least $\log 2/\log(L/\eta)$ corresponding to dissipation concentrated into two curds. However, variants of curdling yield a more relaxed relationship between $D\sim 0$ and concentrates in a few blobs.) We shall see in Section 3 that, among curdling processes of given D, B is greatest in the fractally homogeneous case. Hence, the bound $B\leq 1$ is of wide generality. Sulem & Frisch 1975 were able to rederive it by an entirely different argument from the characteristic that for $D=0$ everything concentrates in a small number of blobs.

2. THE FRACTALLY HOMOGENEOUS CASE WITH OTHER PRESCRIBED DOMAINS Ω.

To discuss Obukhov's specification of Ω further, we shall find it useful (however cumbersome) to decompose this specification into parts of increasing degrees of arbitrariness: a) one should replace ε by the average ε_r of the local dissipation rate $\varepsilon(\mathbf{x})$, taken over an appropriate domain Ω; b) this domain Ω should be independent of $\varepsilon(\mathbf{x})$; c) Ω should be three-dimensional, something like the sphere whose poles are P' and P''. Without going so far as to question assumption a) above, we shall (in Section 5) question both b) and c). In the present Section we shall keep b) and question c). That is, we shall suppose that Ω is fixed but make Ω nearly one-dimensional, namely choose for it a cylinder of radius 2η and axis $P'P''$, or make it strictly one-dimensional, namely (for reasons of symmetry) the segment $P'P''$.

When Ω is $P'P''$, the results are more complex than when Ω is Obukhov's sphere, because in the limit $\eta \to 0$, the probability of $P'P''$ hitting turbulence depends on the value of D. When $D<2$, we know this probability is zero. When $D>2$, we know it to be positive because of the nondegeneracy of linear cross-sections, and it turns out that the expression familiar from Section 1 continues to be valid: the hitting probability is approximately equal to $(r/L)^{3-D}$. As a result, the dependence of $\langle(\Delta \mathbf{u})^2\rangle$ on L/r and of $E(k)$ upon Lk goes as in Section 1, except for a single change, a vital one. Here the coefficients E_0 and $V_{1/3}$ remain positive if $D>2$, but vanish if $D\leq 2$. In particular, the exponent of r is restricted to the narrower range of values between 2/3 and 1, and the exponent of k^{-1} always lies above the Kolmogorov value 5/3, but below the "Burgers" value 2. The latter constitutes the bound corresponding to the stage when all turbulent diffusion *within the segment* $P'P''$ reduces to a few blobs.

3. DISSIPATION GENERATED BY WEIGHTED CURDLING

As was the case for the correlation in Chapter III, the formal argument can be borrowed from Monin & Yaglom 1975, with one exception: just like in Chapter III, one *must not*, and we *shall not*, replace $\log W$ by its Gaussian approximation. The exact result, supposing that $r >> \eta$, is as follows

$$\begin{aligned}\langle \varepsilon_r r^{2/3}\rangle &= V\, r^{2/3}\,[\langle W^{2/3}\rangle]^{\log_\Gamma(L/r)}\\ &= V\, r^{2/3}\,(L/r)^{-B},\end{aligned}$$

with $B = -3f(2/3)$, and

$$E(k) = E_0\, L^{-B}\, k^{-5/3-B}.$$

To evaluate $f(2/3)$, we shall return to the determining function $f(h)$. By convexity, the $0\leq h\leq 1$ portion of the graph of $f(h)$ lies between the h axis and the tangent to $f(h)$ at $h=1$, whose slope is equal to $f'(1) = (3-D)/3$.

As a result, given any value $D<3$, B can range from the maximum value $B = 3f'(1)/3 = (3-D)/3$ (obtained in the fractally homogeneous case) down to 0. (It is possible to show that this last value cannot be attained, but can be approached arbitrarily closely. So it is conceivable, however unlikely, that intermittency should bring no change to the $k^{-5/3}$ spectral density.)

The general inequality $B \leq (3-D)/3$ generalizes the equality $B=(3-D)/3$ valid in the fractally homogeneous case. The value corresponding to a lognormal W is (as known to Kolmogorov 1962) $\mu/9$. Written in terms of $D = 3-\mu/2$, it yields $B = (3-D)/4.5$. This value confirms that the changes in the $2/3$ exponent can be smaller than $(3-D)/3$.

4. THE "BURGERS" THRESHOLD SPECTRUM k^{-2} AND THE DIMENSION $D=2$ ARE RELATED IN ABSOLUTE BUT NOT IN WEIGHTED CURDLING. THIS LAST FACT IS PARADOXICAL.

Formally, the preceding argument is easily generalized to Burgers turbulence and more generally to turbulence with $\Delta u = |P'P''|^{2H}$. The value $B = -3f(2/3)$ is simply replaced by $B = -3f(2H)$. It follows that the Burgers case $H=1/2$, and this case only, has the remarkable property that $B \equiv 0$. The value of the spectral exponent is independent, not only of D but of the random variable W. In other words, even after Burgers turbulence is made intermittent as a result of curdling, its spectral density continues to take the familiar form k^{-2}.

More generally, the "Burgers threshold" will be defined as the point where the intermittency has the intensity needed for the spectrum to become k^{-2}. It is a well-known fact (exploited in Mandelbrot 1975a) that the k^{-2} spectrum prevails when the turbulent velocity change is due to a finite number of two-dimensional shocks of finite strength. Hence it was expected that one should find that the spectrum is k^{-2} in the case of fractal homogeneity with $D=2$. This dimension marks the borderline between the cases when the segment $P'P''$ does, or does not, have a positive probability of hitting dissipation.

On the other hand, it seems that the logical correspondence between $D=2$ and $B=1/3$ fails in the case of weighted curdling. Example: for $D=2$, the lognormal approximation combined with the choice of Obukhov sphere for Ω yields $E(k) = E_0 L^{-2/9} k^{-17/9}$ with $E_0>0$. Even though (assuming it is confirmed that turbulence satisfies $D>2$) the behavior of the spectrum about $D=2$ has no practical effect, the fact that $17/9<2$ constitutes a paradox that must be resolved. We shall postpone this task to Chapter V.

5. NON-PRESCRIBED DOMAINS Ω IN FRACTALLY HOMOGENEOUS TURBULENCE AND THE ISSUE OF TOPOLOGICAL CONNECTEDNESS

Let us resume the discussion of the choice of Ω, started in Section 3. The use of any *fixed* Ω implies the belief that the mutual interaction between $\mathbf{u}(P')$ and $\mathbf{u}(P'')$ is on the average independent of the fluid flow in between the points P' and P''. It is, however, worth at least a brief consideration to envision interactions propagating along lines, say, of least resistance. In the all-or-nothing fractally homogeneous case, it may well be possible to join P' and P'' by a line Λ such that $\int_\Lambda \varepsilon(\mathbf{x})d\mathbf{x}=0$. If so, one may well argue that $\Delta\mathbf{u}$ should vanish.

One is tempted in this spirit to replace ε_r by $(1/r)$ glb $\int_\Lambda$, a short notation for the product of $(1/r)$ by the greatest lower bound of $\int_\Lambda \varepsilon(\mathbf{x})d\mathbf{x}$ along all lines Λ joining P' to P''. The principle of the new specification of Ω is radically different from $\Lambda = P'P''$, because, if accepted, it would open the door to topology. In particular, two of the shapes to be studied in the turbulence Chapter of the English version of *Fractals* (namely, the Sierpiński sponge and pastry shell) have the same $D>2$ but very different topology. For the former glb $\int_\Lambda=0$ for any P' and P'', while for the latter glb $\int_\Lambda=0$ if P' and P'' lie in the same cutout, and glb $\int_\Lambda>0$ otherwise. Since turbulence does in fact exist so that $\langle(\Delta\mathbf{u})^2\rangle>0$, the acceptance of the Λ that minimizes $\int_\Lambda$ would lead to the following tentative inference: Among all sets of two points P' and P'', selected at random under the constraint that $|P'P''|=r$, sets in which every line from P' to P'' hits the carrier of turbulence must have a positive probability. In other words, the probability of P' and P'' being separated by "sheets" of turbulence must be non-vanishing. The mathematical nature of this tentative inference is entirely distinct from the fractal inequality $D>2$; the latter was metric, while the present one combines topology with probability. A mixture of theoretical argument with computer simulations shows there exist a critical dimension D_0, such that the probability of the set generated by absolute curdling being sheet-like is zero when $D<D_0$ and positive when $D>D_0$. This D_0 is much closer to 3 than to 2.

Further, it is tempting to constrain Λ to stay in the Obukhov sphere, and designate the restricted glb by glb$'\int_\Lambda$. If so, the inference that $\langle \text{glb}'\int_\Lambda\rangle>0$ would involve a combination of topological, probabilistic and metric features; this lead can not yet be developed any further.

The preceding reference to topology is extremely tentative, by far less firmly established than the fractal inequality $D>2$. It accepts without question two results of Kolmogorov and Obukhov: the 1941 link between $\langle(\Delta\mathbf{u})^2\rangle$ and a uniform ε, and the 1962 link between $\langle(\Delta\mathbf{u})^2\rangle$ and the expectation of $\varepsilon^{2/3}$. Moreover, the all-or-nothing fractal homogeneity may well be too flimsy a model to support such extensive theorizing.

V. THE INNER SCALE AND THE DISSIPATIVE RANGE.

Thus far the existence of actual dissipation was only acknowledge indirectly, by introducing an inner scale η which, like L, was arbitrarily imposed from the outside. We shall now dig deeper into the classical result of Kolmogorov: Taylor homogeneous turbulence with the viscosity ν and a uniform rate of dissipation ε, the dissipative range is vanishingly narrow around the inverse of $\eta_3 = \nu^{3/4}\varepsilon^{-1/4}$. One reason for the notation η_3 is that the letter η was used up above; a more consequent reason will appear momentarily. One aspect of η_3 is that, if the spectrum $E(k) = E_0 \varepsilon^{2/3} k^{-5/3}$ is truncated at an appropriate numerical multiple of $k = 1/\eta_3$, the relationship $\varepsilon = \nu \int_{1/L<k<1/\eta} k^2 \ E(k)\, dk$ becomes an identity. In this Chapter we shall first examine formally the changes due to intermittency. Then we shall proceed to an actual analysis of the inner scale of curdling. In the fractally homogeneous case, the inner scale will continue to be defined as the inverse of the spectrum's truncation points. This result had already been obtained by Novikov & Stewart, but it deserves a more careful analysis. In all other cases the result is *quite different*. The analysis will show the necessity of a dissipative range that does *not* reduce to the neighborhood of any single value $1/\eta$, but has a definite width determined by the value of $f''(1)$.

1. TRUNCATION POINT FOR THE POWER-LAW SPECTRUM

From Chapter IV, the spectral density of velocity in intermittent turbulence is of the form $E(k) = E_0 \varepsilon^{2/3} k^{-5/3} (Lk)^{-B}$. Suppose we want the relationship $\varepsilon = \nu \int_{1/L<k<1/\eta} k^2 E(k) dk$ to continue as an identity. Then, up to numerical factors, one must have $\eta = \eta'_D$, where η'_D is defined by

$$\varepsilon = \nu\, \varepsilon^{2/3} L^{-B} \eta'_D{}^{-4/3-B}$$

$$\eta'_D = [(\nu^3/\varepsilon)\, L^{-3B}]^{1/(4-3B)}$$

$$\eta'_D/L = (\eta_3/L)^{1/(1-3B/4)}.$$

Since $0 \le B \le 1$, we find that $\eta'_D << \eta_3$. For given D, η'_D is a monotone function of B. Thus, when B reaches its maximum value $B = (3-D)/3$, η'_D reaches its minimum value $L(\eta_3/L)^{4/(D+1)}$ and $1/\eta'_D$ reaches its maximum. Note that, in addition to ν and ε, the value of η'_D depends upon L.

More generally, if one stays within a sub-domain of length scale r, much smaller than L, in which the average dissipation is ε_r, one will have the new inner scale $\eta'_D(r)$ such that

$$\nu^{3/4}\, \varepsilon_r^{-1/4}/r = [\eta'_D(r)/r]^{1-3B/4}.$$

2. CRITIQUE OF AN INNER SCALE OF INTERMITTENT TURBULENCE TENTATIVELY SUGGESTED BY KOLMOGOROV

The need to reexamine the concept of inner scale had already been felt by Kolmogorov (1962). On p. 83 of this work (seventh formula) he suggested for this role the expression $\nu^{3/4}\langle\varepsilon_r\rangle^{-1/4}$ which occurs in the left-hand side of the last formula of the preceding Section. He did not explain his choice, and made no further use of it. Actually, it seems hard to retain. A first odd feature of his definition is that when $r = L$, his modified inner scale reduces to η_3. Hence, contrary to η_D, it is independent of the degree of intermittency. A second odd feature relates to $r \to 0$. To describe it, let us follow Kolmogorov in assuming $\log\varepsilon_r$ to be lognormal, with the variance $\mu \log (L/r)$ and a mean adjusted to insure that $\langle\varepsilon_r\rangle = \varepsilon$. It follows that

$$\langle\varepsilon_r^{-1/4}\rangle = \varepsilon^{-1/4} \exp[(1/2)(-1/4)(-5/4)\mu\log(L/r)]$$
$$= \varepsilon^{-1/4} (L/r)^{5\mu/32}.$$

Hence, as $r \to 0$, the modified Kolmogorov scale *increases* on the average and may exceed r. We shall not attempt to unscramble this concept.

3. INNER SCALE OF CURDLING IN THE FRACTALLY HOMOGENEOUS CASE

The truncation of $E(k)$ shows that the energy cascade must stop when reaching eddies on the order of magnitude of η'_D. But what about the curdling cascade? It too must have an end, to be followed by dissipation. We shall now identify the scale η_D at which it stops.

In the fractally homogeneous case, η_D turns out to be identical to the η_D' defined through $E(k)$. Consider, indeed, a cube of side L filled with a Taylor homogeneous turbulent fluid of viscosity ν, dissipation ε and inner scale η_3. Since we assume that the increasingly small curds created by a Novikov-Stewart cascade are themselves Taylor homogeneous; these curds are endowed with a classical Kolmogorov scale varying with the cascade stage. We assume moreover that the instability and breakdown leading to curdling are encountered if and only if the curd size exceeds the Kolmogorov scale. (This assumption can be seen to be equivalent to a little-noticed condition of Novikov & Stewart, as reported in Monin & Yaglom 1975, p. 611.)

The first curdling stage leads to curds of side L/Γ in which dissipation is equal to either 0 or $\varepsilon\,\Gamma^{3-D}$. In the empty cells, the Kolmogorov scale is infinite, and of course further curdling is impossible. In the first stage curds, the inner scale is $\eta^{(1)} = \eta_3\Gamma^{-(3-D)/4}$. In the m-th stage curds, the average dissipation is $\varepsilon\Gamma^{m(3-D)}$, the curd size is $L\Gamma^{-m}$, and the inner scale is therefore $\eta^{(m)} = \eta_3\,\Gamma^{-m(3-D)/4}$. We see that the inner scale and the curd size both decrease with $1/m$. Our postulate being that there is no further curdling after these two scales meet, we are left with the criterion $\eta_3\,\Gamma^{-m(3-D)/4} \sim \Gamma^{-m}L$, i.e., $\eta_3/L = [\Gamma^{1-(3-D)/4}]^{-m}$. The solution turns out to yield

$\Gamma^{-m} L = \eta_D$, with η_D identical to the η'_D obtained earlier in this Chapter through the truncation of $E(k)$. Hence the fractal dimension rules not only the manner in which Novikov-Stewart curdling proceeds, but the point where it stops. We find, in addition, that it is reasonable to assume that the cutoff of $E(k)$ near $1/\eta_D$ is very sharp.

Digression concerning curdling in spaces of Euclidean dimension $\Delta > 3$. The derivation of η_D has relied on the fact that, in a space of Euclidean dimension $\Delta = 3$, the decrease in η_m is less rapid than the decrease in curd size. However, this last feature is highly dependent upon $\Delta - D$, and therefore upon the value of Δ. As in many other fields of physics, a qualitative change may be observed when $\Delta \neq 3$. Indeed, our stability criterion readily yields the result that a nonvanishing inner scale *need not exist*. It exists if and only if $\Delta - D < 4$. Its value is given by the relation

$$(\eta_\Delta/L) \sim (\eta_D/L)^{1-(\Delta-D)/4}.$$

The necessary and sufficient condition $\Delta - D < 4$ for the existence of a non-vanishing inner scale is peculiar but not very demanding. One amply sufficient condition is $\Delta < 4$. (However, in order that curdling continue forever, meaning $\eta_D = 0$, the converse condition $\Delta > 4$ is only necessary, *not* sufficient.) Another amply sufficient condition for $\eta_D > 0$ is $\Delta - D < 1$, which we know expresses that linear cross-sections are *not* almost surely empty. These various conditions make it clear that a vanishing inner scale can at most be observed for phenomena that are very much sparser than the turbulent dissipation presently under study. Much sparser even than the Leray-Scheffer conjectural singularities of the Navier-Stokes equations.

Nevertheless, odd as the result may be, our criterion does indicate that, when $\Delta - D > 4$, *a curdling cascade will continue forever, without any physical cutoff, even when the viscosity is positive.* I do not know what this result means, and what its implications concerning dissipation are. It seems to be trying to tell us something about the singularities in the ultimate solution of the equations of motion of some physical system, but I cannot guess of which one.

4. INNER SCALE OF CURDLING WHEN INTERMITTENCY IS GENERATED BY WEIGHTED CURDLING FIRST ROLE OF $f''(1)$.

In the case of weighted curdling, as we shall now proceed to show, the inner scale is best studied in two approximations. The first one yields a single typical value. The interesting fact is that this value turns out to be much smaller than the quantity η'_D obtained through the truncation of $E(k)$. The second approximation shows that said typical value is not very significant and that one must deal with a whole statistical distribution. Strictly speaking, the same situation had already prevailed in all-or-nothing curdling leading to fractally homogeneous intermittency, but in that case $\eta^{(m)}$ was simply binomial, equal to either $\eta_3\Gamma^{-m(3-D)/4}$ or infinity and the latter value could be neglected. The same cannot be done in the case of weighted curdling.

Recall that, if the curdling cascade could continue forever, the dissipation density $\varepsilon(\mathbf{x})$ at the point $\mathbf{x}$ would be a product of weights W, one per cascade stage. It can be written in the form $W_{i_1} W_{i_1 i_2} W_{i_1 i_2 i_3} \ldots$, where the real number $0, i_1 i_2 \ldots$ designates $\mathbf{x}$ in the counting base C, and the W's are an infinite sequence of independent random weights. Similarly, $\eta^{(m)}$ will be written simply in the form $\eta_3[\Pi_{1\leq n\leq m} W_n]^{-1/4}$, Curdling will stop when this random $\eta^{(m)}$ first overtakes the nonrandom $\Gamma^{-m}L$. Taking logarithms, we find that m is the first integer where

$$\Sigma_{1\leq n\leq m}[(-1/4)\log W_n + \log\Gamma] = \log[L/\eta_3].$$

After we select a probability distribution for W, the left-hand side of the above expression will define a random walk with nonrandom drift equal to $\langle \log\Gamma - \log W/4 \rangle$ and an absorbing barrier. The above-defined value of m is therefore merely an instant of absorption or, alternatively, of ruin. Absorption will occur almost surely because the drift turns out to be positive (digression: this is so as long as $\Delta - D < 4$).

First approximation. When $L/\eta_3 >> 1$, the drift tends to overwhelm the randomness, and one can approximate m by the value m^* obtained by the rough approximation which consists in replacing the random walk by its expectation. The proper choice of weights in the above expectation is not obvious, but there is room only to state the result without a full justification. Moreover, in order to avoid irrelevant notational complication, we add the assumption that the values w_g of W are discrete with probabilities p_g. Then the proper intrinsic probability of w_g is not given by p_g itself. Rather, it can be shown to take the form $p_g w_g$. Since $\langle W \rangle = 1$, $\Sigma p_g w_g = 1$; therefore the $p_g w_g$ are acceptable as probabilities. Continuing to use $\langle\ \rangle$ to designate expected values under the probabilities p_g, our criterion yields

$$\eta_3/L = [\Gamma(\exp\langle W\log W\rangle)^{-1/4}]^{-m^*}$$

The result stated in the last form turns out to apply also to nondiscrete W's. Since $-\langle W\log_\Gamma W\rangle = D-3$, the definition of η_D reduces formally to that applicable in the fractionally homogeneous case.

Summary of the first approximation. In weighted curdling the order of magnitude of m is the same as in the all-or-nothing curdling having the same value of D, hence of $f'(1)$. In particular, the order of magnitude of $1/\eta_D$ is much *greater* than the $1/\eta'_D$ deduced in Section 1.

Second approximation. The actual values of m scatter around m^*. For fixed W, the scatter increases with L/η_3. For fixed L/η_3, it is useful to define a standard scatter, to be denoted by σm. It is approximately the ratio of two factors. The first is the standard deviation of the sum of $m*$ factors of the form $-\log W/4$. The variance of W is $\langle W\log^2 W\rangle - \langle W\log W\rangle^2$, which happens to be equal to $\log C\, f''(1) = 3\log\Gamma\, f''(1)$. Hence, the first factor is equal to $[3m*\log\Gamma f''(1)]^{1/2}/4 = [3\log(L/\eta_3)/(1-(3-D)/4)]^{1/2}/4$. The second factor is the expected value of $\log\Gamma - \log W/4$, that is $\log\Gamma - f'(1)\log C/4 = \log\Gamma[1-(3-D)/4]$. Combining the two factors, we obtain the two alternative forms

$$\sigma m = (2/\log\Gamma)(1+D)^{-3/2}[3\log(L/\eta_3)]^{1/2}[F''(1)]^{1/2}$$
$$= (3/\log\Gamma)^{1/2}(1+D)^{-1}[m*]^{1/2}[f''(1)]^{1/2}$$

This is the first time that the value of $f''(1)$ enters in the present discussion. The value of $f'(1)$ enters also, through D, but the result is not very sensitive to it.

5. THE DISSIPATIVE RANGE.

The methods used in Chapters III and IV to evaluate exponents and exponent changes only apply to scales for which curdling has a small probability of having stopped, that is, roughly, from $(1/L)$ to $k\sim(1/L)\Gamma^{m*-\sigma m}$. Going towards higher wave numbers, one encounters next the range from $k\sim(1/L)\Gamma^{m*-\sigma m}$ to $k\sim(1/L)\Gamma^{m*+\sigma m}$. Here, some dissipation is likely to occur in a substantial region of our fluid.

Let us make a few more comments on this topic. By the last result of the preceding Section, the width of the dissipative range, measured in units of $\log_\Gamma k$, is proportional to $[f''(1)]^{1/2}$. When $\log W$ is lognormal, $f(h)$ is parabolic and $f''(1)$ is proportional to μ. More generally, unless the distribution of W is very bizarre, one has approximately $f(2/3)\sim f(1) - f'(1)(1/3) + f''(1)(1/3)^2/2$. That is, $(3-D)/3-B \sim f''(1)/6$. This relation holds even if B and $(3-D)/3$ do not bear to each other any numerical relationship of the kind that holds when W is lognormal and $B = (3-D)/4.5$. In other words, the width of the dissipative range – measured on the $\log_\Gamma k$ scale – is typically the square root of the defect of B with respect to the fractally homogeneous approximation.

It was to be expected that each of these quantities should be a monotone increasing function of the other. Indeed, the inequality $1/\eta_D >> 1/\eta'_D$ expresses that the spectrum $k^{-5/3-B}$ relative to the inertial range cannot be extrapolated consistently. The corresponding distribution of energy among the wave numbers decreases very much too slowly as k increases, which implies that the whole energy would be completely exhausted well before reaching $k \sim 1/\eta'_D$. The greater the difference $1/\eta - 1/\eta'_D$, the sooner must this inertial range law $k^{-5/3-B}$ cease to apply.

The expressions that apply in the dissipative range and replace coefficients such as B, will be described elsewhere.

ACKNOWLEDGMENT

Before the present final version, I had the benefit of penetrating comments by Uriel Frisch: by listing a few of the things he did not understand, he motivated me to substantial further development.

REFERENCES

Batchelor, G. K. & Townsend, A. A., 1949. The nature of turbulent motion at high wave numbers. *Proceeding of the Royal Soc. of London* **A 199**, pp. 238-255.

Berger, J. M. & Mandelbrot, B. B., 1963. A new model for the clustering of errors on telephone circuits. *IBM Journal of Research and Development:* **7**, pp. 224-236.

Corrsin, S., 1962. Turbulent dissipation fluctuations. *Physics of Fluids* **5**, pp. 1301-1302.

Feller, W., 1971. *An Introduction to Probability Theory and its Applications* (Vol. 2, 2d ed.) New York: Wiley.

Kahane, J. P., 1974. Sur le modèle de turbulence de Benoit Mandelbrot, *Comptes Rendus* (Paris) **278A**, pp. 621-623.

Kahane, J. P. & Mandelbrot, B. B., 1965. Ensembles de multiplicité aléatoires, *Comptes Rendus* (Paris) **261**, pp. 3931-3933.

Kahane, J. P. & Peyrière, J., 1976. Sur certaines martingales de B. Mandelbrot. *Advances in Mathematics* (in the press).

Kahane, J. P. & Salem, R., 1963. *Ensembles parfaits et séries trigonométriques.* Paris: Hermann.

Kolmogorov, A. N., 1962. A refinement of previous hypotheses concerning the local structure of turbulence in a viscous incompressible fluid at high Reynolds number. *Journal of Fluid Mechanics* **13**, pp. 82-85.

Kuo, A. Y. S. & Corrsin, S. 1972. Experiments on the geometry of the fine structure regions in fully turbulent fluid. *Journal of Fluid Mechanics* **56**, pp. 477-479.

Leray, J., 1934. Sur le mouvement d'un liquide visqueux emplissant l'espace, *Acta Mathematica.* **63**, pp. 193-248.

Mandelbrot, B., 1965. Self-similar error clusters in communications systems and the concept of conditional stationarity. *IEEE Transactions on Communications Technology:* **COM-13**, pp. 71-90.

Mandelbrot, B., 1967. How long is the coast of Britain? Statistical self-similarity and fractional dimension. *Science* **155**, pp. 636-638.

Mandelbrot, B., 1972. Possible refinement of the lognormal hypothesis concerning the distribution of energy dissipation in intermittent turbulence. *Statistical Models and Turbulence.* (ed. Rosenblatt & Van Atta), pp. 333-351. New York: Springer.

Mandelbrot, B., 1974a. Multiplications aléatoires itérées, et distributions invariantes par moyenne pondérée. *Comptes Rendus,* (Paris) **278A**, pp. 289-292 & 355-358.

Mandelbrot, B., 1974b. Intermittent turbulence in self-similar cascades: divergence of high moments and dimension of the carrier. *Journal of Fluid Mechanics* **62**, pp. 331-358.

Mandelbrot, B., 1975a. On the geometry of homogeneous turbulence, with stress on the fractal dimension of the iso-surfaces of scalars. *Journal of Fluid Mechanics.*

Mandelbrot, B., 1975b. *Les objets fractals: forme, hasard et dimension.* Paris & Montreal: Flammarion.

Mandelbrot, B., 1976a. Géométrie fractale de la turbulence. Dimension de Hausdorff, dispersion et nature des singularités du mouvement des fluides. *Comptes Rendus,* (Paris) **282A**, pp.119-120.

Mandelbrot, B., 1976b. *Fractals: form, chance and dimension.*

Mattila, P., 1975. Hausdorff dimension, orthogonal projections and intersections with planes. *Annales Academiae Scientiarum Fennicae* **A I 1**.

Monin, A. S. & Yaglom, A. M., 1975. *Statistical Fluid Mechanics: Mechanics of Turbulence.* Cambridge, Mass.: MIT Press.

Novikov, E. A., 1969. Scale similarity for random fields. *Doklady Akademii Nauk SSSR* **184**, pp. 1072-1075. (English trans. *Soviet Physics Doklady* **14**, pp. 104-107.)

Novikov, E. A. & Stewart, R. W., 1964. Intermittency of turbulence and the spectrum of fluctuations of energy dissipation. *Isvestia Akademii Nauk SSR; Seria Geofizicheskaia* **3**, p. 408.

Obukhov, A. M., 1962. Some specific features of atmospheric turbulence. *Journal of Fluid Mechanics* **13**, pp. 77-81.

Scheffer, V., 1976. Equations de Navier-Stokes et dimension de Hausdorff. *Comptes Rendus* (Paris) **282A,** pp. 121-122.

Sulem, P. L. & Frisch, U., 1975. Bounds on energy flux for finite energy turbulence. *Journal of Fluid Mechanics* **72,** pp. 417-423.

Tennekes, H., 1968. Simple model for the small scale structure of turbulence. *Physics of Fluids,* **11**, pp. 669-672.

Von Neumann, J., 1949-1963. Recent theories of turbulence (a report to ONR) *Collected Works,* **6**, pp. 437-472.

Yaglom, A.M., 1966. The influence of the fluctuation in energy dissipation of the shape of turbulent characteristics in their inertial interval. *Doklady Akademii Nauk SSSR* **16,** pp. 49-52. (English trans. *Soviet Physics Doklady,* **2,** 26-29.)

FIGURE 1. PLANAR FRACTAL OBTAINED BY ABSOLUTE CURDLING.

Random curdling proceeds on a square grid. We show the effect of four stages, each of which begins by dividing the cells of the previous stage into $5^2=25$ subcells, then "erasing" 10 of them to leave the remaining 15 as "curds".

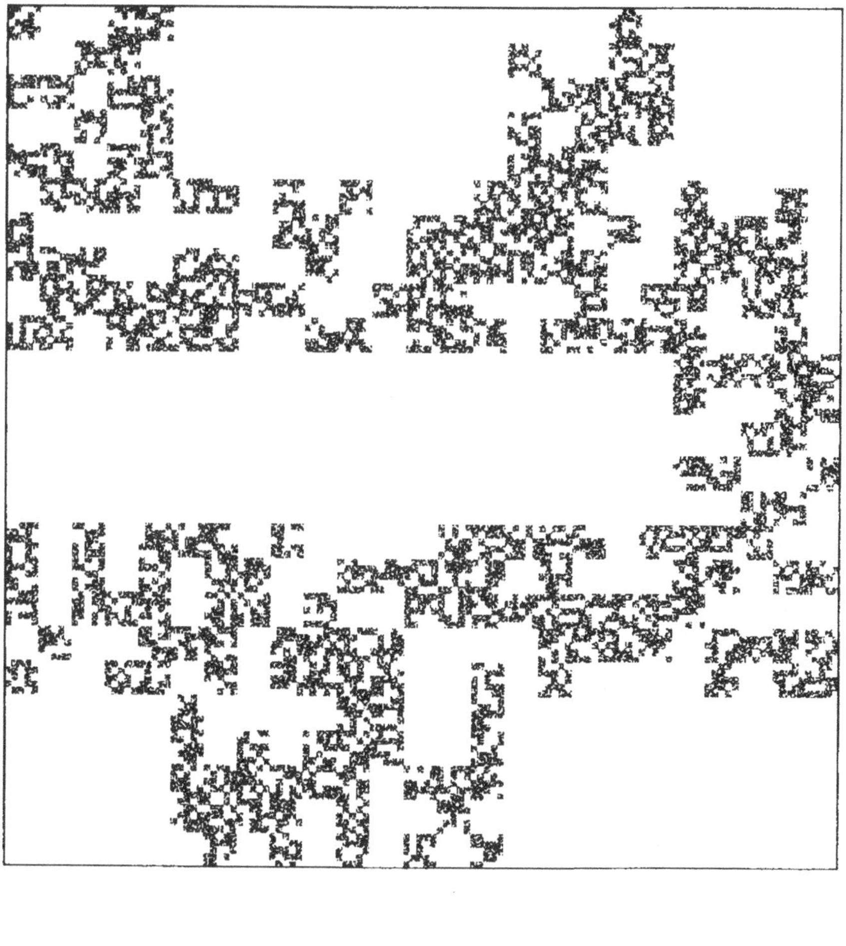

FIGURE 2. THE SELF SIMILARITY DEFINITION OF THE FRACTAL DIMENSION.

A segment of line can be paved by – and is therefore equivalent to – $N=5$ replicas of itself reduced in the ratio $r=1/5$. A square is equivalent to $N=25$ replicas of itself reduced in the ratio $r=1/5$. The same property of self similarity is obviously encountered in the pattern of Figure 1: it is equivalent to $N=15$ replicas of itself reduced in the ratio $r=1/5$. In each of the classical cases, the concept of dimension can be associated with self similarity, and one has $D=\log N/\log(1/r)$. The point of departure of fractal geometry is that this last expression a) remains well defined and b) happens to be useful for all self-similar sets such as the pattern of Figure 1, and that it is not excluded for D to be a fraction.

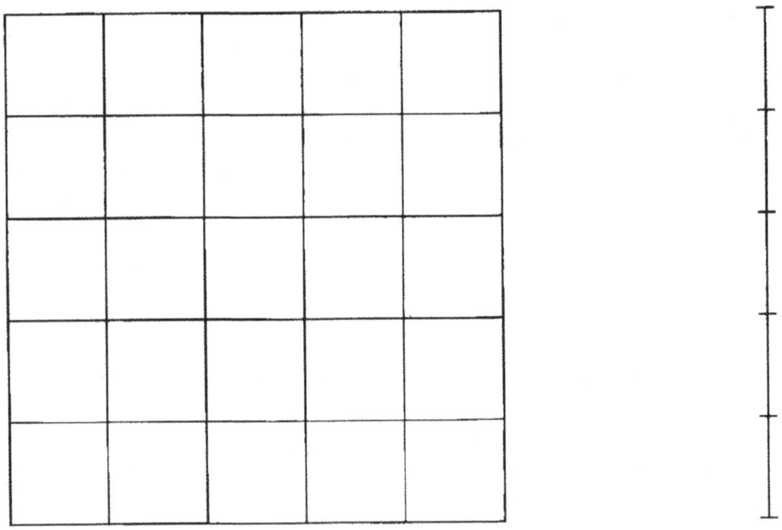

FIGURE 3. THE DETERMINING FUNCTION $f(h)$.

The two lines represent two determining functions $3f(h)$ which yield the same value of $3-D=.45$. The present paper concentrates upon the roles played in the theory of intermittency by the quantities $f(2/3)$, $f(2)$ and $f''(1)$. Earlier work, Mandelbrot 1974a,b, had concentrated on the role played by $f'(1)$ and the α's (the latter are not shown here). The lower line in the present Figure is straight of equation $3f=.45(h-1)$. It corresponds to fractally homogeneous turbulence and is the lowest compatible with the given D. The upper line, which is the parabola $3f=.45h(h-1)$, corresponds to lognormal intermittency with $\mu=.9$. For other forms of curdling, the determining function can lie between the above lines or even higher than the parabola. Two examples are of interest. The fractally homogeneous case can be changed so that the value 0 is replaced by some scatter of values slightly above it, while the value of $1/p$ is slightly changed to keep D invariant. Alternatively, the lognormal can be truncated sharply. In either case, the resulting line $f(h)$ will be approximately straight for abscissas to the right of $h=1$ and approximately parabolic to the left.

The concept of approximation used in the bulk of probability theory is of little value in the present context. A random variable W' may be a close approximation to W and still lead to a markedly different determining function and hence to a markedly different form of intermittency.

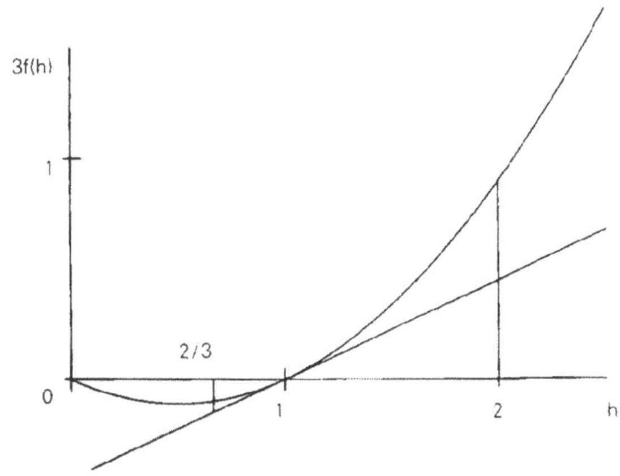

THE LORENZ ATTRACTOR AND THE PROBLEM OF TURBULENCE

David RUELLE *

I.H.E.S. 35 route de Chartres 91440 Bures sur Yvette

I - Introduction.

Turbulence in the flow of liquids is a fascinating phenomenon. This may partly explain the conceptual confusion which exists in the scientific literature as to the nature of this phenomenon.

My own opinion, and that of some other people, is that turbulence at low Reynolds numbers corresponds to a mathematical phenomenon observed in the study of solutions of differential equations

$$dx/dt = X(x) \quad . \tag{1}$$

The equation just written has to be understood as a time evolution equation in several dimensions. The mathematical phenomenon referred to is that in many cases, solutions of (1) have an asymptotic behavior when $t \longrightarrow \infty$ which appears erratic, chaotic, "turbulent", and the solutions depend in a sensitive manner on initial condition.

According to conventional ideas, when the Reynold's number of a fluid is increased a number of independant frequencies, or "oscillations", successively appear in the fluid. The superposition of a sufficiently large number of frequencies would produce turbulence. This view was developed by E. Hopf [4], who explained the occurence of various frequencies by successive "bifurcations". A more popular version of the same ideas is due to Landau [6]. By a "superposition of k frequencies" is here meant a quasi periodic time dependence of the form

$$x(t) = F(\omega_1 t, \ldots, \omega_k t) \tag{2}$$

where F is periodic of period 2π in each argument separately, and the ω_i are k irrationally related frequencies. Such a quasi periodic motion does appear erratic, chaotic, or turbulent, but does not depend in a sensitive manner on initial condition. F. Takens and myself [11] noted however that for $k \geq 4$, a small perturbation could transform a quasi periodic motion into one that would again look "turbulent", and

---

* David Ruelle, n'ayant pu participer aux Journées, nous reproduisons avec l'autorisation de l'Auteur et des Editeurs, le texte de sa Communication à la Conference "Quantum Dynamics Models and Mathematics", Bielefeld, Septembre 1975.

would furthermore have a sensitive dependence on initial conditions. This can be interpreted in physical terms by saying that putting a nonlinear coupling between 4 or more oscillations can produce a "turbulent" time evolution with sensitive dependence on initial condition. Technically we proved that "strange Axiom A attractors" could occur. Axiom A flows and attractors have been introduced by Smale, and are very nice mathematical objects to work with from an abstract point of view. Instead of discussing those I shall however turn now to the discussion of an object which has greater geometric simplicity, and intuitive appeal : the Lorentz attractor.

Lorenz' work [8], which was unfortunately overlooked on [11], is the first attempt at interpreting turbulence by solutions of differential equations which appear chaotic, and have sensitive dependence on initial condition. The ideas of Lorenz and those of Takens and myself have recently received support from the theoretical work of Mc Laughlin and Martin [9] and the experimental work of Gollub and Swinney [3]. It can now be hoped that more experimental results on the onset of turbulence in various systems will become available in the next few years ; their theoretical interpretation will constitute a worthy challenge for the mathematical physicist.

II - The Lorenz Attractor.

2.1. Generalities. The Lorenz equations are the following

$$\begin{aligned} \dot{x} &= -\sigma x + \sigma y \\ \dot{y} &= -xz + rx - y \qquad (3) \\ \dot{z} &= xy - bz \end{aligned}$$

where $\dot{x} = dx/dt$, etc... and σ, r, b are positive numbers. These equations are obtained as an approximation to partial differential equations describing convexion in a fluid layer heated below (Bénard problem). The unknown functions in the Bénard problem are expanded in Fourier series and an infinite system of coupled differential equations is obtained. By putting all Fourier coefficients equal to zero except three, one gets the truncated system (3). The system (3) is of course a badly mutilated version of the original postial differential equations, and one may well wonder what relation its solutions have to the original problem. Things may not be as bad as they appear, but we won't go into that question. We shall interest ourselves here only in mathematical phenomena exhibited by solutions of (3) : their chaotic appearance and sensitive dependence on initial condition. These phenomena occur for a suitable range of values of the parameters σ, r, b, and Lorenz made for his numerical studies the choice

$$\sigma = 10 \;, \quad b = 8/3 \;, \quad r = 28 \;. \qquad (4)$$

2.2. *There is a bounded region* B *of* $\mathbb{R}^3$ *such that every solution of* (3) *eventually becomes trapped in* B .

Lorenz notices that this holds for a general system of the form

$$dx_i/dt = \sum_{j,k} a_{ijk}\, x_j\, x_k - \sum_j b_{ij}\, x_j + c_i$$

where $\sum a_{ijk}\, x_i x_j x_k$ vanishes identially and $\sum b_{ij} x_i x_j$ is positive definite. One verifies indeed immediately that

$$\frac{1}{2}\frac{d}{dt}\sum_i x_i^2 = -\sum_{ij} b_{ij}\, x_i\, x_j + \text{lower order}$$

is negative when $\sum_i x_i^2$ is large enough. The system (3) satisfies the above assumptions after the change of variables

$$x' = x\ ,\quad y' = y\ ,\quad z' = z - r - \sigma\ .$$

2.3. *The time evolution given by* (3) *contracts volumes in* $\mathbb{R}^3$ *at a constant rate.*

$$\frac{\partial \dot{x}}{\partial x} + \frac{\partial \dot{y}}{\partial y} + \frac{\partial \dot{z}}{\partial z} = -\ (\sigma+b+1)\ .$$

Notice that for the values (4) of the parameters this is $-\ 13^{2/3}$, which is a very fast rate.

2.4. *The system* (3) *is invariant under the symmetry.*

$$(x,\ y,\ z) \longrightarrow (-x,\ -y,\ z)$$

2.5. *Steady state solutions and bifurcations.*

Clearly, the right-hand side of (3) vanishes at the point $\mathcal{O} = (0,\ 0,\ 0)$, which is thus a steady state solution. For $r<1$ this is the only steady state solution and it is attracting. This is seen by looking at the matrix of partial derivatives of the right-hand side of (3) at the point $\mathcal{O}$:

$$\left(\frac{\partial x_i}{\partial x_j}\right) = \begin{pmatrix} -\sigma & \sigma & \\ r & -1 & \\ & & -b \end{pmatrix}.$$

The eigenvalues of this matrix have negative real part (are in fact real negative), showing that $\mathcal{O}$ is attracting.

When r becomes larger than 1, O loses its attracting character (one eigenvalue becomes positive) and two new steady state solutions appear :

$$C = (\sqrt{b(r-1)}\ ,\ \sqrt{b(r-1)}\ ,\ r-1)$$

$$C' = (-\sqrt{b(r-1)}\ ,\ -\sqrt{b(r-1)}\ ,\ r-1)$$

O, C, C' are the only steady state solutions for $r>1$. The bifurcation which leads to the creation of C, C' is illustrated in Fig.1. This picture shows that C, C' should be attracting. Looking at the matrix of partial derivatives $(\partial x_i/\partial x_j)$ at C or C' confirms this : the eigenvalues have a negative real part : first they are real negative, then one pair becomes complex conjugate. However if

$$(5) \qquad r > \sigma(\sigma+b+3)(\sigma-b-1)^{-1} > 0$$

then the complex conjugate pair has now a positive real part. The values (4) satisfy (5). We have thus a situation where there are three steady state solutions, none of them attracting. Let us examine the situation in detail.

a) near O points come in along a two-dimensional surface (stable manifold of O) and go out along a line (unstable manifold).

b) near C points come in along a line and go out along a two-dimensional surface ; they spiral out because the eigenvalues with positive real part have a non vanishing imaginary part.

c) similarly near C'.

In this situation, numerical calculations by Lorenz showed that the solution of (3) have an apparently erratic, chaotic, or turbulent behavior.

2.6. <u>The Poincaré map</u>.

To understand what is happening, it is convenient to go from 3 to 2 dimensions, by using a Poincaré map. Let P be a point in the plane $z = 27$ containing C and C'. We assume that $dz/dt < 0$ at P, i.e. P is inside an equilateral hyperbola through C and C'. We follow the integral curve through P, and let ϕP be the next point at which this curve crosses the plane $z = 27$, going downwards. The map $\phi : P \longrightarrow \phi P$ is our Poincaré map. Oscar Lanford put the problem to a computer, and Fig.2 is the answer. It is seen that the successive images of a point P by the Poincaré map ϕ tend to lie on two arcs Γ and Γ' *). The points C, C' are respectively on the continuation of Γ and Γ'. Also drawn is the line $\sum$ of points

---

* The point P itself might lie anywhere. It has therefore not been drawn, neither have the first few points $\phi^k P$.

which do not come back to cross the plane $z = 27$, because the integral curve they determine goes to O (Σ is part of the unstable manifold of O).

Because of the symmetry 2.4, it suffices to understand the action of ϕ on Γ. Since points close to C are spiraling away from C, it is understandable that the piece of Γ contained between C and Σ is stretched in the manner described by Fig.3. The computer tells us that the piece of Γ beyond Σ is mapped onto Γ' as indicated in Fig.4. Notice that, using the notation of Fig.3, 4, ϕB is not defined, and if $x \to \Sigma$ then ϕx tends to B' or B'' depending on whether x is on the same side of Σ as C, or on the other side*).

2.7. Γ and Γ' are not line arcs.

I must apologize for having implied that Γ and Γ' are line arcs. A little thinking shows that this would be in contradiction with the uniqueness of solutions of differential equations. To see better what happens, it is convenient to use the symmetry 2.4 and identify symmetric points. The map ϕ is now well defined at B since B' and B'' are identified. Figs 3, 4, become Fig.5. This describes the image under ϕ of a line arc approximating Γ. It so happens that the two line arcs on the picture of the right are practically on top of each other. This is not too astonishing in view of 2.3. Altogether one expects that when $n \to \infty$, the points $\phi^n P$ tend to sets Γ and Γ' which look like Cantor sets in cross section.

2.8. Sensitive dependence on initial condition.

It should now be clear that the map ϕ compresses $\Gamma \cup \Gamma'$ in one direction and stretches it in another. Therefore given two points P and P' close together, in general the stretching will have the effect that $\phi^n P$ and $\phi^n P'$ will be farther and farther apart. In fact, their distance will increase exponentially with n, as long at least as it is not too large. Better than that (or worse than that) after a while $\phi^n P$ and $\phi^n P'$ will fall on different sides of Σ, and from then on their futures are totally dissimilar.

Going back to the original system (3) we see that its solutions depend sensitively on initial condition. This would not be true for a quasi-periodic motion (2) as one readily checks.

2.9. Hyperbolicity.

The combination of compressing in one direction and stretching in another is called <u>hyperbolicity</u>. We refrain from a formal definition. A little thinking shows that hyperbolicity causes sensitive dependence on initial condition. This is exactly what happens for Axiom A time evolutions, which are defined by a hyperbolicity

---

* The points B', B'' are on the unstable manifold of O, cf. section 2.5(a).

requirement. Technically however, the Lorenz equations do not satisfy Axiom A . That is because the integral curves can come arbitrarily close to 0 , and are then "slowed down" for an arbitrarily long time : this spoils some uniformity in the hyperbolicity required for Axiom A to hold.

2.10. The work of Lanford and Guckenheimer.

The facts mentioned above about the solutions of (3) have been obtained numerically by use of a computer by Lorenz and by Lanford. Unfortunately (fortunately for mathematicians) computers do not yet prove theorems. Therefore I must make the reservation that, while one is quite confident that things are as described above, there are no proofs yet, and proofs may be hard to obtain. Work in this direction is being done by Lanford* and Guckenheimer**. Fig.5 suggests the study of maps of a line segment onto itself of the type Fig.6. Such maps have been investigated by Li and Yorke [7], and W. Thurston (unpublished). From this work one can derive information on periodic orbits of (3).

2.11. Hysteresis.

Notice that the sets Γ and Γ' in Fig.2 do not extend all the way to C and C' respectively. This is understandable since ϕ actually pushes Γ away from C, and the piece of Γ near C has to be re-fed every time from Γ' . When the parameter r is decreased it is thus possible that $\Gamma \cup \Gamma'$ will remain an attracting set for ϕ , but that C and C' also become attracting. Depending on the initial condition the system will thus be in $\Gamma \cup \Gamma'$, or in C , or in C' . A corresponding situation will prevail for the system (3) for a certain range of r. In fact the system will be near a different attracting set depending on whether r is raised from low values or decreased from high values. This phenomenon is called hysteresis. In the case at hand, I do not expect hysteresis to be important, due to the smallness of the gap between Γ and C .

III - Questions.

An attractor for a flow (or differential equation) is a compact set Λ such that all points sufficiently close to Λ tend to Λ under time evolution when the time tends to $+\infty$. To this definition some "irreducibility" condition should be added ; for Axiom A flows it is required that Λ be connected. The following striking result holds.

---

* O. Lanford, private communication.

** R. Bowen, private communication.

3.1. Theorem.

Let $t \longrightarrow x(t)$ *be the time-evolution for a* C^2 *Axiom A flow. Then for almost all initial condition* $x(0)$ *with respect to Lebesgue measure* *, $x(t)$ *tends to an attractor*** *when* $t \longrightarrow +\infty$.

The open set of those $x(0)$ *such that* $x(t)$ *tends to a given attractor* Λ *is called the basin of* Λ.

There is a probability measure μ *with support* Λ *such that for almost all* $x(0)$ *with respect to Lebesgue measure in the basin of* Λ, *and every continuous function* ϕ *on this basin*

$$\lim_{T\to\infty} \frac{1}{T}\int_0^T \phi(x(t))dt = \int \mu(dy)\ \phi(y) . \tag{6}$$

The measure μ *is invariant and ergodic under time evolution.*

It is a natural question whether this carries over to the Lorenz attractor. My guess is that it does. Notice that (6) looks like the ergodic theorem but the formula holds for almost all $x(0)$ with respect to Lebesgue measure, and the mearure μ in the right-hand side is not Lebesgue measure. Theorem 3.1 is proved in [1], which also contains a characterization of μ by a variational principle. Would this variational principle apply to the Lorenz attractor ?***

If (6) holds one can define a time correlation function

$$F(t) = \int \mu(dx(0))\ \phi(x(0))\ \phi(x(t)) - (\int \mu(dx)\ \phi(x))^2$$

where ϕ is assumed to be differentiable.

3.2. Conjecture. For the Lorenz attractor, $F(t)$ tends to zero exponentially fast when $t \longrightarrow \infty$.

The corresponding fact is known to be true in the discrete-time Axiom A case (Axiom A diffeomorphisms [10]). It expresses the sensitive dependence on initial condition (or the loss of information on initial condition). In any case one expects that $F(t)$ tends to zero, even if not exponentially (mixing).

IV - Some Misconceptions.

There is now detailed experimental evidence on the onset of turbulence.

---

* We call Lebesgue measure the measure defined by a Riemann metric, it is thus not unique, but sets of Lebesgue measure zero are unambiguously defined.

** An Axiom A flow has a finite number of attractors.

*** If it does, this permits an estimate of the entropy of the flow.

Gollub and Swinney [3] have indeed studied an example of transition to turbulence with the following features :

a) The transition is sharp and without hysteresis (within experimental error ; there is probably some hysteresis).

b) The nature of the correlation functions changes abruptly*. The oscillations predicted by the quasi-periodic Landau picture are not present.

These results and the analysis of Mc Laughlin and Martin [9] indicate that the ideas of Lorenz [8], Takens and myself [11] are in better agreement with experiment than traditional interpretations of turbulence.

In view of all this I want to discuss critically some ideas, which appear frequently in the literature, and which I feel to be misconceptions.

4.1. That an external source of "random noise" is necessary to explain the apparent loss of information in turbulent flow.

The sensitive dependence on initial condition seems to explain all what has to be explained. Of course there is some noise in experiments. There are now theorems to the effect that for Axiom A systems a little noise only changes a little bit the measure μ of (6) [see Sinai [12], Kifer [5]]. Whether the effect of the noise will be weak or important will depend on whether the attractor one is on is strongly or weakly attracting**).

4.2. That a measure describing turbulence should be approximately Gaussian.

This idea seems to stem from the notion that in turbulent motion a large number of different frequencies of small amplitude are superposed incoherently. This comes from the Landau picture of turbulence. A moment's thinking shows however that according to the Landau picture different frequencies are in different directions of the "phase space" of the system (functional space of velocity fields), and don't therefore superpose incoherently. On the other hand it is clear that the Lorenz attractor does not carry an approximately Gaussian measure (cf. Fig.2). Also the asymptotic measure μ on an Axiom A attractor defined by (6) cannot be a approximately Gaussian because one can prove [1] that Axiom A attractors have Lebesgue measure zero.

4.3. That the condition of stationarity is more or less sufficient to determine the measure describing turbulence.

---

* The frequency spectrum shows an abrupt transition from a regime with sharp spikes, to one with only continuous spectrum.

** Near a normal bifurcation the attracting character of an attractor is weak, and therefore "noise amplification" should occur.

Let μ be a measure describing the statistical, or time-averaged, behavior of a turbulent flow (cf. equation (6)) - then μ is of course invariant under time evolution. However, on a strange* Axiom A attractor there is a continuum of different probability measures invariant under time evolution. To fish the right one out one needs some extra information which can, in the Axiom A case, be given by a variational principle (cf. [1]).

To summarize, I think it would be a miracle if the usual procedure of imposing stationarity, truncating the resulting system of equations, and looking for a Gaussian solution, would lead to results much related to physics.

V - Outlook.

There remains a lot to be done to have a good understanding of the onset of turbulence. Let me mention two problems which obviously deserve investigation.

5.1. Study of asymptotic measures on general attractors.

The conjecture is that for almost all initial points with respect to Lebesgue measure the time average (6) makes sense (at least for almost all time evolutions - in some sense). The problem would be to study the measures μ thus obtained. How generally do the time correlation functions decrease exponentially ?

Notice that the existence of (6) for almost all initial points is proved for Axiom A flows [1], flows asymptotic to a torus with a quasi-periodic motion on it (because of unique ergodicity), and Hamiltonian flows on compact manifolds (this follows from the invariance of the Liouville-Lebesgue measure and the ergodic theorem -- I am indebted to D.S. Ornstein for this remark).

5.2. Study of bifurcations occurring in concrete problems at the onset of turbulence.

This is a difficult but exciting problem. The only existing work in this direction is that of McLaughlin and Martin[9],which is excellent although perhaps a bit over enthusiastic. As new experimental work becomes available (like the beautiful recent work of Gollub and Swinney [3]), we realize our lack of information on obvious theoretical questions. For instance, in spite of the work of Lorenz and Lanford, simple questions about the Lorentz attractor are still unanswered : how much hysteresis is there ? What is the frequency spectrum like ? Also what are simple attractors present in low dimension apart from the Lorenz attractor ?

In conclusion let me express my feeling that, after decades of misconceptions,

---

* We call an Axiom A attractor "strange" if it does not reduce to a fixed point or a closed orbit of the flow.

we are beginning to have correct ideas on the time dependence in turbulence near its onset. This solves the problem of turbulence only in part, however. In particular the space dependence has striking features (see [2]) which will probably require new conceptual ideas for their elucidation.

Acknowledgments

This work was performed during the summer of 1975, while the author was visiting the Institute for Advanced Study, Princeton University, and Yeshiva University. Thanks are due to S. Adler, A.S. Wightman, and J.L. Lebowitz for these invitations. I am also very indebted to O. Lanford whose work on the Lorenz attractor I have freely used in the present report.

References

[1] R. Bowen and D. Ruelle - The ergodic theory of Axiom A flows. Inventiones math. To appear.

[2] R.P. Feynman, R.B. Leighton, and M. Sands - The Feynman lectures on physics. 2. Addison-Wesley, Reading, Mass., 1964.

[3] J.P. Gollub and H. Swinney - Onset of turbulence in a rotating fluid. Preprint.

[4] E. Hopf - A mathematical example displaying features of turbulence. Commun. Pure appl Math. 1, 303-322 (1948).

[5] Kifer

[6] L.D. Landau and E.M. Lifshitz - Fluid Mechanics. Pergamon, Oxford, 1959.

[7] T.Y. Li and J. Yorke - Period three implies chaos. Preprint.

[8] E.N. Lorenz - Deterministic nonperiodic flow. J. atmos. Sci. 20, 130-141 (1963).

[9] J.B. Mc Laughlin and P.C. Martin - Transition to turbulence of a statically stressed fluid system. Phys. Rev. Lett. 33, 1189-1192 (1974) Phys. Rev. A. 12, 186-203 (1975).

[10] D. Ruelle - A measure associated with Axiom A attractors. Amer. J. Math. To appear.

[11] D. Ruelle and F. Takens - On the nature of turbulence. Commun. Math. Phys. 20, 167-192 (1971) ; 23, 343-344 (1971).

[12] Ia. G. Sinai - Gibbs measures in ergodic theory. Russ. Math. Surveys 166, 21-69 (1972).

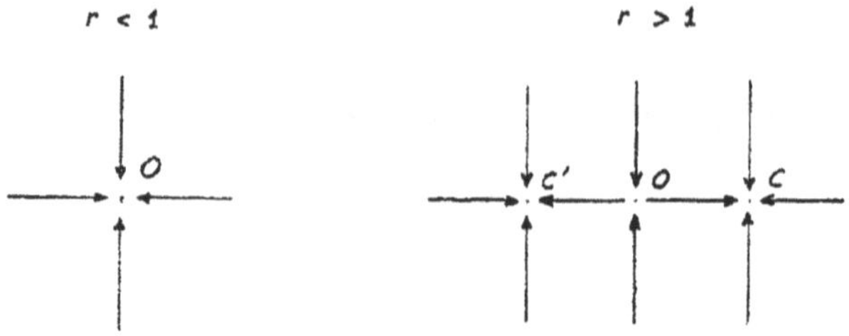

Figure 1.

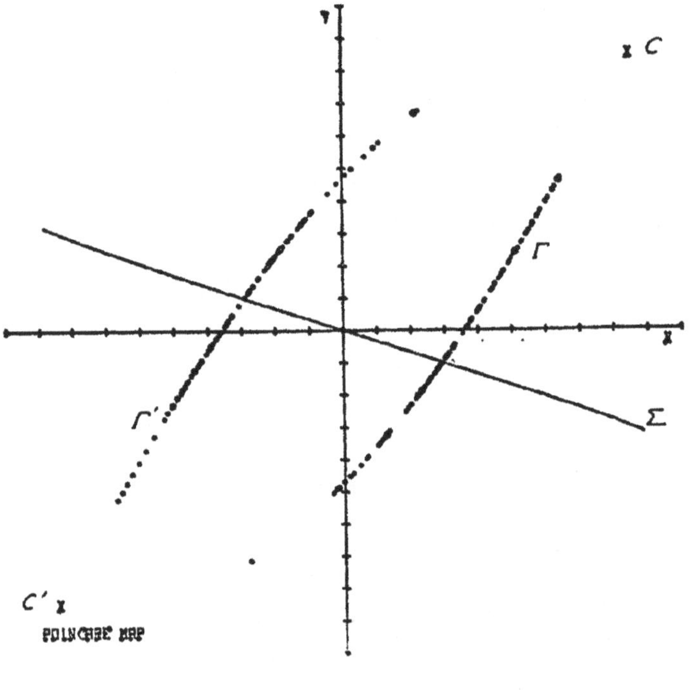

Figure 2.

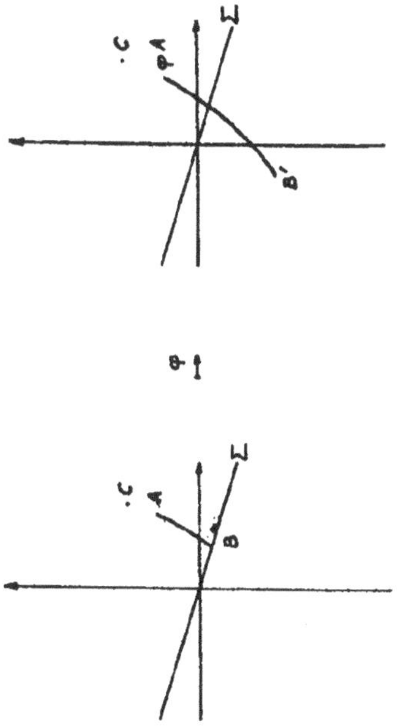

Figure 3.

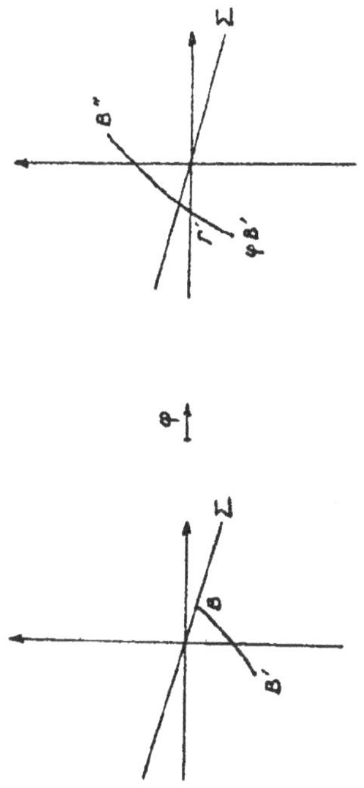

Figure 4.

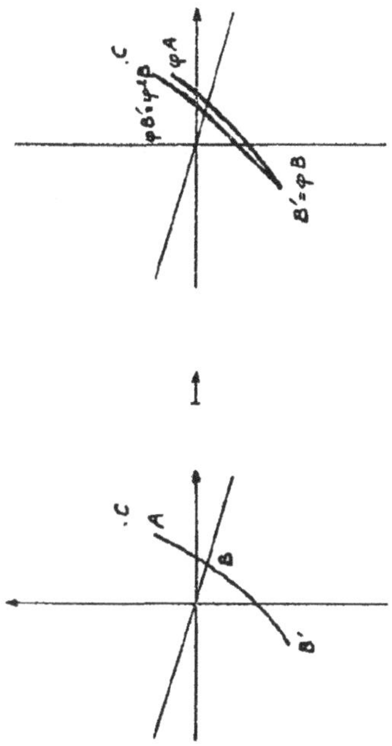

Figure 5.

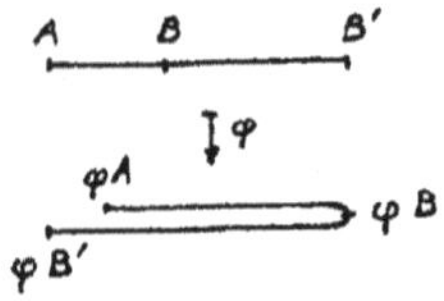

Figure 6.

Pattern Formation in Convective Phenomena

by

D.H. Sattinger
Minneapolis, Minnesota USA

Of particular interest in fluid mechanics is the formation of flow patterns at the onset of instability, or, as it is sometimes called, "symmetry breaking instabilities". The appearance of convection cells in the Bénard problem [13], or the Taylor vortices in the Taylor problem [7], constitute well-known classical examples of such symmetry breaking instabilities. Other situations of interest are the formation of patterns in a spherical geometry, as in astrophysical or geophysical applications [2], [9]. In the case of the Bénard problem the solution prior to the onset of convection is invariant under the entire group of rigid motions; but after the onset of instability the fluid motion arranges itself in cellular, often hexagonal, patterns. Such flow patterns are invariant only under a crystallographic subgroup of the group of rigid motions. (See the plate in the article by L.A. Segel [13].) While the mechanisms governing the <u>onset</u> of the instability are by now well understood, the mechanisms governing the <u>selection</u> of the particular pattern are not.

Owing to the covariance of the above problems (as well as many others) group theoretic methods can be brought to bear on them with considerable profitability. A full discussion of group-theoretic methods and their application to the Bénard problem will appear in a forthcoming article [12]. In this talk I shall summarize the principle group-theoretic ideas and describe some of the results which may be obtained with their aid.

Consider a physical system which is described by some set of equations which may be written generally as

$$(1) \qquad G(\lambda, u) = 0$$

where u is an element of a Banach space $\mathcal{B}$ and G is a non-linear (analytic) mapping from $\mathcal{B}$ to another Banach space $\mathcal{F}$. Let $\mathcal{G}$ be a group and suppose T_g is a representation of $\mathcal{G}$ on both $\mathcal{B}$ and $\mathcal{F}$. That is, $T_g : \mathcal{B} \to \mathcal{B}$, $T_g : \mathcal{F} \to \mathcal{F}$, and $T_{g_1 g_2} = T_{g_1} T_{g_2}$. We say that (1) is covariant with respect to T_g if

$$(2) \qquad T_g\, G(\lambda, u) = G(\lambda, T_g u) \quad .$$

For example, in the case of the Bénard problem (1) is a system of nonlinear partial differential equations, the Boussinesq equations; and $\mathcal{B}$ consists of a vector-valued function, $u = \mathrm{col}(u_1(\underline{x}), u_2(\underline{x}), u_3(\underline{x}), \theta(\underline{x}), p(\underline{x}))$, where u_1, u_2, u_3 are the Cartesian components of the velocity, θ is the temperature, and p is the pressure. If $g = \{r, a\}$ is a rigid motion, say $g\,\underline{x} = r\underline{x} + a$, where r is a rotation and a is a vector, the representation of $\mathcal{G}$ (the group of rigid motions) on $\mathcal{B}$ is given by

$$(3) \qquad (T_g u)(\underline{x}) = \begin{bmatrix} r & 0 & \\ 0 & 1 & \\ & & 1 \end{bmatrix} \begin{bmatrix} u_1 \\ u_2 \\ u_3 \\ \theta \\ p \end{bmatrix} (g^{-1}\underline{x})$$

The covariance of the Boussinesq equations is easily established along the same lines as that of the Navier Stokes equations [6], [11].

The appearance of cellular solutions can be described mathematically as the bifurcation of doubly periodic solutions. Let Λ be a lattice of vectors in the plane: If $\underline{\omega}$, $\underline{\omega}' \in \Lambda$ then $m\underline{\omega} + n\underline{\omega}' \in \Lambda$ for any integers m and n . A function u is doubly periodic if $u(\underline{x} + \underline{\omega}) = u(\underline{x})$ for all $\underline{\omega} \in \Lambda$. We shall say that u is Λ - periodic if we wish to specify a particular lattice. We denote by $\mathcal{S}(\Lambda)$ the subgroup of rigid motions which leaves the class of Λ - periodic functions invariant.

In the case of the Bénard problem we construct a Banach space of doubly periodic vector-valued functions as follows. We let $u(\underline{x}) = \text{col}\ (u_1,\ldots,u_5)$ where $u_4 = \theta$ and $u_5 = p$. The point $\underline{x} = (x,y,z)$ lies in $-\infty < x,\ y < \infty$, $0 < z < d$. We then consider u as a mapping from R^2 to the space of vector valued functions $f(z) = \text{col}\ (f_1(z),\ldots,f_5(z))$. Suitable norms are the Hölder norms or the Dirichlet norms. The Boussinesq equations comprise an elliptic system of partial differential equations in the sense of Agmon, Douglis, and Nirenberg (see Fife [4]) and the Frechet derivative satisfies a Fredholm alternative on the space of doubly periodic functions.

If the temperature and pressure are measured from their rest state values then the conduction solution of (1) is simply $u = 0$. To determine the bifurcating convection solutions of (1) we apply the methods of bifurcation theory. Let the parameter λ be chosen so that the critical parameter value is $\lambda_z = 0$. We denote $G_u(0,0)$ by L_0 and by $\mathcal{N}(\Lambda)$ the kernel of L_0 in the space of Λ- periodic functions.

<u>Theorem 1</u>: $\mathcal{N}(\Lambda)$ <u>is finite dimensional and invariant under</u> $\mathcal{G}(\Lambda)$.

<u>Proof</u>: $\mathcal{N}(\Lambda)$ is finite dimensional because the operator $G_u(0,0)$ is an elliptic Fredholm operator. $\mathcal{N}(\Lambda)$ is invariant under T_g since $[L_0,T_g] = 0$. This relationship is proved as follows: Since $G(\lambda,u)$ is covariant, $T_g\, G(\lambda,u) = G(\lambda,T_g u)$, and so $T_g\, G_u(\lambda,u) = G_u(\lambda,T_g u)T_g$. Setting $u = \lambda = 0$, the result follows.

Therefore, T_g restricted to $\mathcal{N}(\Lambda)$ is a finite dimensional representation of $\mathcal{G}(\Lambda)$. We shall see later that $\mathcal{N}(\Lambda)$ is irreducible for appropriate choices of Λ .

By the Lyapounov-Schmidt procedure the bifurcation problem is reduced to solving a system of bifurcation equations

$$(4) \qquad F(\lambda,v) = 0$$

where $v \in \mathcal{N}(\Lambda)$. If we introduce a basis $[\varphi_1,\dots,\varphi_n]$ of $\mathcal{N}(\Lambda)$ we may write (4) as a system of n equations in n unknowns: $F_i(\lambda,z_1,\dots,z_n) = 0$, $i = 1,\dots,n$.

<u>Theorem 2</u>: <u>The covariance of the original equations is inherited by the bifurcation equations</u> (4). <u>That is</u> $T_g\, F(\lambda,v) = F(\lambda,T_g v)$.

Theorem 2 is immediate upon examining the Lyapounov-Schmidt procedure. (See [11] for details.) Theorem 2 may be used to determine the explicit structure of the bifurcation equations up to some undetermined constants which must be evaluated in terms of the physical constants in the given problem. That is, the structure of the bifurcation equations can be determined from the group-theoretic arguments alone, and is independent of the structure of the equations (1).

Let us describe the subspace $\mathcal{N}(\Lambda)$. To simplify the discussion we work with scalar valued functions $u(\underline{x})$ where $\underline{x} \in R^2$. The algebra is the same as in the case of the Bénard problem (see [12]). Since $\mathcal{N}(\Lambda)$ is invariant under the full group of translations we choose a basis for $\mathcal{N}(\Lambda)$ of the form $\{e^{i\langle \underline{\omega}_j, \underline{x}\rangle}, j = 1,2,\dots\}$ where the vectors $\underline{\omega}_1, \underline{\omega}_2, \dots$ lie in the plane and $\langle\ ,\ \rangle$ is the Euclidean inner product. The functions $e^{i\langle \underline{\omega}_j, \underline{x}\rangle}$ are the one-dimensional irreducible representations of translation group and are known as the Bloch functions in solid state physics. Since $e^{i\langle \underline{\omega}_j, \underline{x}\rangle}$ is Λ periodic we must have $e^{i\langle \underline{\omega}_j, \underline{x}+\underline{\omega}\rangle} = e^{i\langle \underline{\omega}_j, \underline{\omega}\rangle} e^{i\langle \underline{\omega}_j, \underline{x}\rangle}$ for all $\underline{\omega} \in \Lambda$. Hence $e^{i\langle \underline{\omega}_j, \underline{\omega}\rangle} = 1$ for all $\underline{\omega} \in \Lambda$ and the vectors $\underline{\omega}_j$ must lie in the dual lattice, Λ'. The dual lattice is constructed by choosing a basis dual to that of Λ: If Λ is generated by $\underline{\omega}_1$ and $\underline{\omega}_2$ choose vectors $\underline{\omega}_1'$, $\underline{\omega}_2'$ such that $\langle \underline{\omega}_i', \underline{\omega}_j\rangle = 2\pi\delta_{ij}$, $i = 1,2$. Then Λ' is the lattice generated by $\underline{\omega}_1'$ and $\underline{\omega}_2'$.

Now suppose $e^{i\langle \underline{\omega}', \underline{x}\rangle}$ belongs to $\mathcal{N}(\Lambda)$. If r is a rotation-reflection which leaves $\mathcal{N}(\Lambda)$ invariant we must have $e^{i\langle \underline{\omega}', r^{-1}(\underline{x}+\underline{\omega})\rangle} = e^{i\langle r\underline{\omega}', \underline{\omega}\rangle} e^{i\langle r\underline{\omega}', \underline{x}\rangle} = e^{i\langle \underline{\omega}', r^{-1}\underline{x}\rangle}$ for all $\underline{x}$. Consequently we must have $r\underline{\omega}'$ in the dual lattice as well. We denote the group of rotations and reflections which leave Λ - periodic functions invariant by $\mathcal{D}(\Lambda)$. There are three types of lattices in the plane generated by basic vectors $\underline{\omega}_1$ and $\underline{\omega}_2$ of equal length. (From now on we drop the primes and use $\underline{\omega}_1$ and $\underline{\omega}_2$ to denote the basic vectors of the dual lattice.) These are pictured below in Figure 1. One could also consider lattices where $\underline{\omega}_1$ and $\underline{\omega}_2$ were not both of the critical wavelength, but heuristic considerations (see [6], [12]) suggest such classes of solutions would not be stable.

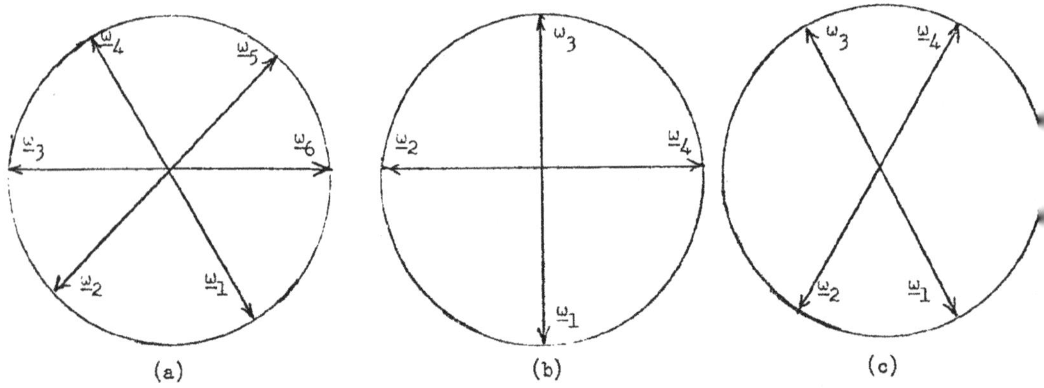

Fig. 1

The kernel $\mathcal{N}(\Lambda)$ consists of the wave functions $\psi_j(\underline{x}) = e^{i\langle\underline{\omega}_j,\underline{x}\rangle}$ where the vectors $\underline{\omega}_j$ lie on the circle of critical wave vectors. (The radius of this circle, that is $\sqrt{\langle\underline{\omega}_1,\underline{\omega}_1\rangle}$, is determined in the process of investigating the onset of instability. See [12].) Let $\psi_1(\underline{x}) = e^{i\langle\underline{\omega}_1,\underline{x}\rangle}$. The other basis vectors of $\mathcal{N}(\Lambda)$ are obtained by operating on ψ_1 by the rotations and reflections in $\mathcal{D}(\Lambda)$. The vectors $\underline{\omega}_j$ are the vertices of a regular polygon, which we shall call the fundamental polygon for the lattice Λ.

Theorem 3: Let T_g be the representation of $\mathcal{S}(\Lambda)$ on $\mathcal{N}(\Lambda)$. The action of T_g is as follows:

(i) If Ta is a translation $Ta\,\psi_j = e^{i\langle\underline{\omega}_j,\underline{a}\rangle}\psi_j$

(ii) If g is a rotation-reflection in $\mathcal{D}(\Lambda)$ T_g acts as a permutation on the ψ_j, permuting them in the same way as g permutes the vertices of the fundamental polygon.

(iii) $T_g\psi_j = \overline{\psi}_j$ if $g\,\underline{\omega}_j = -\underline{\omega}_j$.

Proof: (i) $(T_a\psi_j)(\underline{x}) = e^{i\langle\underline{\omega}_j,\underline{x}+\underline{a}\rangle} = e^{i\langle\underline{\omega}_j,\underline{a}\rangle}\psi_j(\underline{x})$.

(ii) $T_r\psi_j(\underline{x}) = e^{i\langle\underline{\omega}_j,r^{-1}\underline{x}\rangle} = e^{i\langle r\underline{\omega}_j,\underline{x}\rangle} = \psi_{r(j)}$.

(iii) If $r\underline{\omega}_j = -\omega_j$ then $T_r\,e^{i\langle\underline{\omega}_j,\underline{x}\rangle} = \overline{\psi}_j$.

For example, suppose Λ is the hexagonal lattice of Fig. 1(a). The symmetry group of the hexagon is generated by the permutations $\alpha = (123456)$ and $\beta = (26)(35)$. The corresponding elements of $\mathcal{D}(\Lambda)$ are a rotation through 60° and a reflection across the $\underline{\omega}_1 - \underline{\omega}_4$ axis. Therefore $T(\alpha)\psi_1 = \psi_2$, $T(\alpha)\psi_2 = \psi_3$, $\ldots T(\alpha)\psi_6 = \psi_1$, while $T(\beta)\psi_1 = \psi_1$, $T(\beta)\psi_2 = \psi_6$, $\ldots$.

If w is a vector in $\mathcal{N}(\Lambda)$ we write

$$w = \sum_j z_j \psi_j ;$$

w is real iff $z_j = \bar{z}_k$ whenever $\psi_j = \bar{\psi}_k$. We have

Theorem 4: For the hexagonal lattice the action of $\mathcal{S}(\Lambda)$ on the components $(z_1,\ldots,z_6)$ is as follows:

(i) $\alpha(z_1,\ldots,z_6) = (z_6,z_1,z_2,\ldots,z_5)$

(ii) $\beta(z_1,\ldots,z_6) = (z_1,z_6,z_5,z_4\cdot z_3,z_2)$

(iii) $T_a(z_1,\ldots,z_6) = (e^{i\langle \underline{\omega}_1,\underline{u}\rangle} z_1,\ldots, e^{i\langle \underline{\omega}_6,\underline{u}\rangle} z_6)$

Furthermore, for real vectors we require $z_4 = \bar{z}_1, z_5 = \bar{z}_2, z_6 = \bar{z}_3$.

Theorem 4 follows from Theorem 3 and the simple observation that the components $z_1\ldots$ and the basis vectors $\psi_1\ldots$ must transform contragradiently.

Using theorem 4 we can compute the general mapping $F(z)$ which is covariant relative to $\mathcal{S}(\Lambda)$. From $T(\alpha)F(z) = F(T(\alpha)z)$ we get $F_6(z_1,\ldots,z_6) = F_1(z_6,z_1,\ldots,z_5)$, and so forth. So once $F_1(z_1,\ldots,z_6)$ is determined the other components of the mapping can be found by cyclic permutation of the arguments. From part ii of Theorem 4 we get

$$(5) \qquad e^{i\langle \underline{\omega}_1,\underline{a}\rangle} F_1(z_1,\ldots,z_6) = F_1(e^{i\langle \underline{\omega}_1,\underline{a}\rangle} z_1,\ldots, e^{i\langle \underline{\omega}_6,\underline{a}\rangle} z_6) .$$

We break F_1 into linear, quadratic, cubic terms. The only linear term satisfying (5) is easily seen to be $F_1(z_1,\ldots,z_6) = a\, z_1$. Then from cyclic permutation we get $F_j(z_1,\ldots,z_6) = a\, z_j$. Thus the only linear covariant mapping is a scalar multiple of the identity.

<u>Theorem 5</u>: <u>$\eta(\wedge)$ is irreducible under</u> T_g .

Theorem 5 is an immediate consequence of Schur's theorem [14]. When F_1 is homogeneous of degree 2 we get from (5)

$$e^{i\langle\underline{\omega}_1,\underline{a}\rangle} z_j z_k = e^{i\langle\underline{\omega}_j,\underline{a}\rangle} z_j \, e^{i\langle\underline{\omega}_k,\underline{a}\rangle} z_k \, .$$

For this to hold for all $\underline{a}$ we must have

$$\underline{\omega}_1 = \underline{\omega}_j + \underline{\omega}_k \, .$$

The only possible solution is $\underline{\omega}_2 + \underline{\omega}_6 = \underline{\omega}_1$, so the general quadratic term is $F_1 = b\, z_2 z_6$, and hence

$$F_2 = b z_3 z_1 \, , \quad F_3 = b z_4 z_2, \ldots \, .$$

In the case of cubic mappings we are led by similar arguments to the condition

$$\underline{\omega}_1 = \underline{\omega}_j + \underline{\omega}_k + \underline{\omega}_\ell \, .$$

The only solutions possible are $\underline{\omega}_j = \underline{\omega}_1$ (say) and $\underline{\omega}_k + \underline{\omega}_\ell = 0$. Thus the possible cubic covariant terms are

$$z_1^2 z_4 \, , \; z_2 z_5 z_1 \, , \; z_3 z_6 z_1 \, .$$

The covariance with respect to β implies that F_1 must be symmetric in $\{z_2, z_6\}$ and $\{z_3, z_5\}$. Therefore the general cubic term is generated by

$$F_1(z_1,\ldots,z_6) = cz_1^2z_4 + dz_1(z_2z_5+z_3z_6) .$$

By similar arguments one could construct covariant terms of arbitrary degree, but it is sufficient to know only the lowest order non-vanishing nonlinear term in order to resolve the bifurcation problem.

The <u>reduced bifurcation equations</u> (that is, the equations consisting of the linear plus lowest order nonlinear terms) can now be written down explicitly. From the fact that the equation (1) is invariant under complex conjugation we get $\overline{F_j(z)} = F_j(\bar{z})$. Also, from $T(\alpha^3)F = FT(\alpha^3)$ we get $F_4(z_1,\ldots,z_6) = F_1(z_4,z_5,z_6,z_1,z_2,z_3)$, etc. Therefore

$$F_4(z_1,z_2,z_3,\bar{z}_1,\bar{z}_2,\bar{z}_3) = \overline{F_1(z_1,z_2,z_3,\bar{z}_1,\bar{z}_2,\bar{z}_3)},\ldots .$$

Therefore the number of equations is reduced from six to three.

These remaining equations are further simplified when one introduces "action-angle" variables

$$z_j = x_j\, e^{i\theta_j} , \quad z_{j+3} = x_j\, e^{-i\theta_j} , \quad j = 1,2,3$$

where the x_j are real and positive. The reduced bifurcation equations are then

$$(6) \qquad \begin{aligned} \lambda x_1 &= bx_2x_3 \\ \lambda x_2 &= bx_3x_1 \\ \lambda x_3 &= bx_1x_2 \end{aligned}$$

if the leading nonlinear terms are quadratic and, if they are cubic, to

$$(7) \qquad \begin{aligned} x_1(\lambda + c(x_2^{\,2}+x_3^{\,2}) + dx_1^{\,2}) &= 0 \\ x_2(\lambda + c(x_3^{\,2}+x_1^{\,2}) + dx_2^{\,2}) &= 0 \\ x_3(\lambda + c(x_1^{\,2}+x_2^{\,2}) + dx_3^{\,2}) &= 0 \,. \end{aligned}$$

A complete analysis of these equations is entirely a matter of straight-forward algebra [12]. In case (6) the bifurcation is transcritical (occurs on both sides of criticality) and the bifurcating solutions are unstable on both sides of criticality. The unique solution to (6) is $x_1=x_2=x_3=-\lambda/b$ if $\lambda/b < 0$ and $x_1=-x_2=x_3=\lambda/b$ if $\lambda/b > 0$.

The case (7) is more complex. Bifurcation may occur above or below criticality depending on the sign of d, $c+d$, or $2c+d$. There are three classes of solutions:

(i) $x_2=x_3=0,\ x_1^{\,2} = -\lambda/d$

(ii) $x_3=0,\ x_1^{\,2}=x_2^{\,2}= -\lambda/(c+d)$

(iii) $x_1^{\,2}=x_2^{\,2}=x_3^{\,2}= -\lambda\ 2c+d)$.

In case (i) the solutions are called <u>rolls;</u> they appear above criticality if $d < 0$ and below criticality if $d > 0$. The rolls are stable to disturbances in the hexagonal lattice if $c < d$; however, they are always unstable in the square and rectangular lattice. Similar considerations apply to the second class. They do not seem to have been discovered before; but, at any rate, they are never stable in the vicinity of the bifurcation point. Finally, the third class is the well-known case of purely hexagonal solutions. The stability of these solutions canno be entirely determined at the lowest order. It can be shown [12] that the stability of the branching solutions of (7) is determined by the eigenvalues of the Jacobian of the reduced bifurcation equations. In the present case the Jacobian of (7) evaluated at the solution in (iii) is

$$\frac{-2\lambda}{2c+d} \begin{bmatrix} d & c & c \\ c & d & c \\ c & c & d \end{bmatrix}$$

and the eigenvalues of this matrix are

$$-2\lambda, \quad \frac{-2\lambda(d-c)}{2c+d}, \quad \frac{-2\lambda(d-c)}{2c+d} .$$

The eigenvalues of the full Jacobian (recall the number of equations was reduced from six to three) are those above plus zero repeated three times. Thus necessary conditions for the stability of the hexagonal solutions are that $\lambda > 0$ (supercritical bifurcation, hence $2c+d < 0$) and $d-c < 0$. These conditions, however, are not sufficient due to the presence of the three zero eigenvalues. Two of them are to be expected and reflect the translational invariance of the original problem. The third arises from a peculiarity in the bifurcation equations: the reduced bifurcation equations (7) are invariant under a three-parameter group - one more than that to which they are entitled by virtue of the two dimensional translational invariance of the original problem. The behavior of this third eigenvalue under the perturbation from the branch point ultimately determines the stability of the hexagonal solutions, but an analysis of this behavior requires a computation of the higher order terms. So the stability of the hexagonal solutions cannot be resolved at the lowest order of the analysis.

Concluding Remarks

We have considered the bifurcation of doubly periodic disturbances for problems covariant with respect to the group of rigid motions. The stability of a bifurcating solution at a branch point is considered relative to disturbances within the same lattice class. Our results show that in general the stability of a particular solution depends on the relative sizes of two parameters which in turn depend on the physical constants of the original problem. In the Bénard problem these parameters include, besides the Rayleigh number, the Prandtl number, the coefficient of surface tension, and the coefficients of buoyancy (relating temperature and density). Thus the stability of a given pattern depends on the competing effects of surface tension, buoyancy, and conduction.

While the onset of convection is well understood, the mechanisms of pattern selection are much more difficult to explain. The results here apply to an infinite plane layer of fluid. This is the approximation usually made when one is interested in the behavior of the fluid in the interior of the region, away from the boundary. When the depth of the fluid is small compared with the diameter of the container, this is a good approximation insofar as the prediction of the onset of convection and of the critical wave number is concerned. Nevertheless, the mathematical task remains open of fitting the cellular solutions to a problem in a finite container. This requires matching the doubly periodic solutions to boundary layer solutions near the walls. Assuming this could be done, the question then remaining is whether the plane layer model adequately predicts the mechanisms of pattern selection. In particular, is a unique pattern selected on the basis of the stability criteria discussed above? Or does the shape of the container ultimately play a role in the selection of a pattern?

References

1. F. Busse, "The Stability of finite amplitude cellular convection and its relation to an extremum principle", Jour. Fluid Mech. 30, (1967), 625-650.

2. _______, "Patterns of Convection in Spherical Shells", Jour. Fluid Mech. (1975), 72, 67-85.

3. J. Birman, "Symmetry changes, phase transitions, and ferroelectricity", FERROELECTRICITY, Elsevier, Amsterdam, 1967.

4. P. Fife, "The Bénard problem for general fluid dynamical equations and remarks on the Boussinesq equations", Indiana Univ. Math. Jour. 20, (1970), pp. 303-326.

5. K. Kirchgässner, "Instability Phenomena in Fluid Mechanics", SYNSPADE (Ed. Hubbard) Academic Press, 1975.

6. K. Kirchgässner and H. Kielhöfer, "Stability and bifurcation in fluid mechanics", Rocky Mountain Journal of Mathematics 3 (1973), 275-318.

7. K. Kirchgässner and P. Sorger, "Branching Analysis for the Taylor problem", Quart. Jour. Mech. and Appl. Math. 22 (1969), 183-210.

8. J.G. Kirkwood and E. Monroe, "Statistical mechanics of fusion", Jour. Chem. Phys. 9 (1941).

9. N. Lebovitz, "Convective Instability in stars", p. 199, Non-equilibrium Thermodynamics, Donnelly, Herman, and Prigogine, Eds. Univer of Chicago Press, 1965.

10. H.J. Raveche and C.A. Stuart, "Towards a molecular theory of freezing", Jour. Chem. Phys. 63 (1975), 1094-1111.

11. D.H. Sattinger, "Group representation theory and branch points of nonlinear functional equations", to appear, SIAM Jour. Math. Analysis.

12. D.H. Sattinger, "Group representation theory, bifurcation theory and pattern formation", preprint.

13. L.A. Segel, "Nonlinear problems in hydrodynamic stability", Non-equilibrium Thermodynamics, Univ. of Chicago Press, 1965.

14. E. Wigner, Group theory and its application to the quantum mechanics of atomic spectra, Academic Press, New York, 1959.

TURBULENCE AND HAUSDORFF DIMENSION

Vladimir Scheffer

Dept. of Mathematics

Stanford University

Stanford, California 94305/USA

§1. Introduction.

In this paper we will prove two theorems relating Hausdorff dimension to the behavior of weak solutions to the Navier-Stokes equations. This work was motivated by the research of B. Mandelbrot [M].

Let $u : R^3 \times R^+ \longrightarrow R^3$ be a "solution turbulente" of the Navier-Stokes equations of viscous three dimensional incompressible fluid flow as defined by J. Leray in [LJ, pp. 240, 241, 235]. The set R^+, defined by $R^+ = \{t: t \geq 0\}$, represents time, the set R^3 represents the space where the fluid moves, and $u(x,t)$ is the velocity of the fluid at the point x and time t. Leray proved in [LJ] that such solutions exist for all initial conditions with finite kinetic energy (this energy condition being $\int_{R^3} |u(x,0)|^2\, dx < \infty$). In the same paper he also proved a regularity theorem for solutions with finite initial kinetic energy. The following rephrasing of his theorem is taken from [S2].

Leray's Theorem. There exists a finite or countable sequence J_0, J_1, J_2, ... such that $J_q \subset R^+$, $J_0 = \{t: t > a\}$ for some a, J_q is an open interval for $q > 0$, the J_q are disjointed, the Lebesgue measure of $R^+ - \bigcup_{q\geq 0} J_q$ is zero, u can be modified on a set of Lebesgue measure zero so that its restriction to each $R^3 \times J_q$ becomes smooth, and

$$\Sigma_{q>0}\ (\mathrm{length}(J_q))^{1/2}$$

is finite.

We will assume that the modification of u on a set of measure zero has been done. Most of the paper is devoted to proving Theorem 1.

Theorem 1. Let u be a "solution turbulente" with finite initial kinetic energy such that the initial conditions are smooth. Let $T > 0$ be given and set $A = \{x \in R^3$: the restriction of u to $\{x\} \times ([0,T] \cap (\bigcup_{q\geq 0} J_q))$ is a bounded function$\}$. Then the Hausdorff dimension of $R^3 - A$ is at most 5/2.

Hausdorff dimension is defined in §2 and we set $[0,T] = \{t: 0 \leq t \leq T\}$. Recall that $\bigcup_{q\geq 0} J_q$ is almost all of R^+. In §3 we will prove Theorem 2 under the hypotheses of Leray's Theorem.

Theorem 2. The 1/2 dimensional Hausdorff measure of $R^+ - \bigcup_{q\geq 0} J_q$ is zero.

Again, the definition of Hausdorff measure can be found in §2. Theorem 2 is essentially the first theorem announced in [S1].

We pause to establish our notation. If a and b are real numbers with $a < b$ then we set $[a,b] = \{t: a \leq t \leq b\}$. If $x \in R^3$ and $r > 0$ then $B(x,r)$ is $\{y \in R^3 : |x - y| \leq r\}$. Here, as in other contexts, "$|\ |$" denotes the euclidean norm. The three components of u will be u_1, u_2, u_3 and the partial derivative of u_i with respect to x_j, x_j being the jth component of R^3, will be written $u_{i,j}$. Iterated partial derivatives will be written $\frac{\partial^2}{\partial x_j \partial x_k} u_i = u_{i,jk}$, etc. A similar notation is used for partial derivatives of functions defined on R^3. The differential of u with respect to the R^3 variables only will be denoted Du. Thus the components of Du are $u_{i,j}$ for $i,j \in \{1,2,3\}$. The differential of a function f defined on R^3 is also denoted by Df. Repeated indices within a term mean that the term is the sum over all values of that index. Thus we have divergence(u) $= \Sigma_{i=1}^{3} u_{i,i} = u_{i,i}$.

For future use, we list two immediate consequences of the definition of "solution turbulente" with finite initial energy (see (1.6) and (1.7) of [S2]).

(1.1) There is a finite constant M such that we have

$$\int_{R^3 \times R^+} |Du|^2 \leq M \quad \text{and} \quad \int_{R^3} |u(x,t)|^2\, dx \leq M \text{ if } t \in \bigcup_{q\geq 0} J_q ;$$

(1.2) divergence(u)$(x,t) = 0$ if $t \in \bigcup_{q\geq 0} J_q$.

An absolute constant is a finite positive constant which does not depend on any of the variables that we will consider except T (see Theorem 1) and M (see (1.1)). The symbol C will always be an absolute constant, and C may have several different values in a single computation. For example, we may have $2C \le C$. This means that one can go through the paper and substitute an expression in T and M for each occurrence of C. This convention, which we take from a similar convention in [S2], shortens the paper.

§2. Hausdorff measure and dimension.

The basic facts about Hausdorff measure can be found in [F, pp. 169-171]. We include the definitions for convenience.

Let S be a subset of R^n for some n, and let a be any real number satisfying $0 \le a < \infty$. We define $C_a = \Gamma(1/2)^a/\Gamma((a/2)+1)$ (Γ is the Gamma function). For any $\delta > 0$ we let $\phi_\delta^a(S)$ be the infimum of all real numbers which can be written in the form $C_a(\Sigma_{k=1}^{\infty}(\text{diameter}(S_k))^a)$ where $S_k \subset R^n$, the diameter of S_k is at most δ, and $S \subset \bigcup_{k=1}^{\infty} S_k$ (the diameter of a set B is the supremum of $\{|x-y| : x \text{ and } y \text{ are elements of } B\}$). The a dimensional Hausdorff measure of S, written $\mathcal{H}^a(S)$, is defined to be the supremum of all numbers $\phi_\delta^a(S)$ as δ ranges over the positive reals. Whenever a is an integer and S is an a dimensional submanifold of R^n, $\mathcal{H}^a(S)$ coincides with the classical a dimensional area of S. We are interested in the case where a is not necessarily an integer. It can be shown (see [F, p. 204]) that for any $0 \le a \le n$ there exists a compact set $S \subset R^n$ such that $\mathcal{H}^a(S)$ is positive and finite.

It can also be shown [F, p. 171] that for any set $S \subset R^n$ there exists $0 \le a \le n$ satisfying

$$\mathcal{H}^b(S) = \infty \text{ if } b < a, \quad \mathcal{H}^b(S) = 0 \text{ if } b > a.$$

This unique number a is called the Hausdorff dimension of S.

We will require the following covering theorem (proved in [F, pp. 141-152]), which we state for dimension 3.

Lemma 2.1. There exists an absolute constant K, which is an integer, satisfying the following property: If Z is a collection of closed spherical balls in R^3 with bounded diameters then there exist sets Z_i, $i = 1, 2, \ldots, K$ such that 1), 2), and 3) hold:

1) $Z_i \subset Z$.

2) *For each i, Z_i consists of disjoint balls.*

3) *Every center of a ball in Z is contained in some ball from the set $\bigcup_{i=1}^{K} Z_i$.*

§3. Proof of Theorem 2.

Let $\delta > 0$ be given. Using Leray's Theorem, we can find n so that $\Sigma_{q=n+1}^{\infty} (\text{length}(J_q))^{1/2} \leq \delta$ and $\Sigma_{q=n+1}^{\infty} \text{length}(J_q) \leq \delta$ hold. Let $K_1, K_2, \ldots, K_m$ be the closed intervals remaining in R^+ after $J_0, J_1, J_2, \ldots, J_n$ have been removed. There exists a partition $\{n+1, n+2, n+3, \ldots\} = I_1 \cup I_2 \cup \ldots \cup I_m$ such that $\bigcup_{q \in I_i} J_q$ contains almost all of K_i. Now

$$\begin{aligned}\Sigma_{i=1}^{m} (\text{length}(K_i))^{1/2} &= \Sigma_{i=1}^{m} (\Sigma_{q \in I_i} \text{length}(J_q))^{1/2} \\ &\leq \Sigma_{i=1}^{m} \Sigma_{q \in I_i} (\text{length}(J_q))^{1/2} \\ &= \Sigma_{q=n+1}^{\infty} (\text{length}(J_q))^{1/2} \leq \delta\end{aligned}$$

and $\text{length}(K_i) = \Sigma_{q \in I_i} \text{length}(J_q) \leq \delta$. Since $\bigcup_{i=1}^{m} K_i$ contains $R^+ - \bigcup_{q=0}^{\infty} J_q$, the conclusion follows from the definition in §2.

§4. L^p inequalities.

Lemma 4.1. There exists an absolute constant C such that

$$(4.1) \quad \int_{R^3} |f|^6 \leq C(\int_{R^3} |Df|^2)^3$$

holds whenever $f: R^3 \longrightarrow R$ is a smooth function that satisfies $f \in L^2$ and $Df \in L^2$.

Proof. In [LO, p. 12] it is proved that (4.1) holds whenever f is smooth with compact support. To prove the general case, we let $\varepsilon > 0$ be given and we choose $r_0 > 0$ sufficiently large so that the inequality

$$\int_{R^3 - B(0,r_0)} |f|^2 \leq \varepsilon$$

holds. Let $r > r_0$ and let $g: R^3 \longrightarrow [0,1]$ be a smooth function such that $g(x) = 1$ for $x \in B(0,r)$, $g(x) = 0$ for $x \notin B(0,r+1)$, and the supremum of $\{|Dg(x)|: x \in R^3\}$ is at most 2. Since fg is smooth with compact support, we can apply the special case to obtain

$$\begin{aligned}\int_{B(0,r)} |f|^6 &\leq \int_{R^3} |fg|^6 \\ &\leq C(\int_{R^3} |D(fg)|^2)^3 \\ &\leq C(\int_{R^3} |f|^2|Dg|^2)^3 + C(\int_{R^3} |Df|^2|g|^2)^3 \\ &\leq C(\int_{R^3 - B(0,r)} |f|^2)^3 + C(\int_{R^3} |Df|^2)^3 \\ &\leq C\varepsilon^3 + C(\int_{R^3} |Df|^2)^3.\end{aligned}$$

Since r can be made arbitrarily large, we conclude

$$\int_{R^3} |f|^6 \leq C\varepsilon^3 + C(\int_{R^3} |Df|^2)^3.$$

Now (4.1) follows by making ε arbitrarily small.

Lemma 4.2. *For each* p *satisfying* $2 < p < 10/3$ *there exists an absolute constant* $C(p)$ *such that*

$$\int_{R^3} |f|^p \leq C(p)(\int_{R^3} |f|^2)^{(6-p)/(10-3p)} + \int_{R^3} |Df|^2$$

holds whenever $f: R^3 \longrightarrow R$ *is a smooth function satisfying* $f \in L^2$ *and* $Df \in L^2$.

Proof. We fix p. For any given f, there are two possibilities: Either (4.2) or (4.3) holds:

(4.2) $\int |f|^p \leq \int |Df|^2$

(4.3) $\int |Df|^2 \leq \int |f|^p$.

If (4.2) holds we are done. We assume that (4.3) holds. Using Hölder's inequality, Lemma 4.1, and (4.3) we obtain

(4.4) $$\begin{aligned}\int |f|^p &= \int |f|^{(3p-6)/2}|f|^{(6-p)/2} \\ &\leq (\int (|f|^{(3p-6)/2})^{4/(p-2)})^{(p-2)/4}(\int (|f|^{(6-p)/2})^{4/(6-p)})^{(6-p)/4} \\ &= (\int |f|^6)^{(p-2)/4}(\int |f|^2)^{(6-p)/4} \\ &\leq C^{(p-2)/4}(\int |Df|^2)^{(3p-6)/4}(\int |f|^2)^{(6-p)/4}\end{aligned}$$

$\le C^{(p-2)/4}(\int |f|^p)^{(3p-6)/4}(\int |f|^2)^{(6-p)/4}.$

The third line in (4.4), Lemma 4.1, and the hypotheses $f \in L^2$, $Df \in L^2$ imply that $\int |f|^p$ is finite. We may assume without loss of generality that $\int |f|^p$ is not zero. Hence we may divide both sides of (4.4) by $(\int |f|^p)^{(3p-6)/4}$ and obtain

(4.5) $(\int |f|^p)^{(10-3p)/4} \le C^{(p-2)/4}(\int |f|^2)^{(6-p)/4}.$

Raising (4.5) to the $(3p - 6)/(10 - 3p)$ power, we obtain

(4.6) $(\int |f|^p)^{(3p-6)/4}$

$$\le C^{((p-2)/4)((3p-6)/(10-3p))}(\int |f|^2)^{((6-p)/4)((3p-6)/(10-3p))}.$$

Substitution of (4.6) into (4.4) yields

$$\int |f|^p \le C(p)(\int |f|^2)^{(6-p)/(10-3p)}.$$

Hence the conclusion of Lemma 4.2 follows even if (4.3) holds.

<u>Lemma 4.3.</u> <u>If</u> $2 < p < 10/3$ <u>then</u> $\int_{R^3 \times [0,T]} |u|^p$ <u>is finite</u>.

Proof. Recall that T is the fixed number in the hypotheses of Theorem 1. Using the fact that $R^+ - \bigcup_{q\ge 0} J_q$ has Lebesgue measure zero, (1.1), the smoothness of u on $R^3 \times J_q$, and Lemma 4.2, we obtain

$$\int_{R^3 \times [0,T]} |u|^p = \int_0^T \int_{R^3} |u(x,t)|^p \,dx\,dt$$

$$\le \int_0^T C(p)(\int_{R^3} |u(x,t)|^2 \,dx)^{(6-p)/(10-3p)} \,dt$$

$$+ \int_0^T \int_{R^3} |Du(x,t)|^2 \,dx\,dt$$

$$\le C(p)\int_0^T M^{(6-p)/(10-3p)} \,dt + M < \infty.$$

<u>Lemma 4.4.</u> <u>Let</u> $2 < p < 10/3$, $\varepsilon > 0$, <u>and</u> $a = (p + 10 + 2p\varepsilon)/(p+2)$. <u>If</u> $x \in R^3$ <u>and</u> n <u>is an integer satisfying</u>

$$\int_{B(x,2^{-n}) \times [0,T]} |u|^p \le 2^{-na} \quad \text{and}$$

$$\int_{B(x,2^{-n}) \times [0,T]} |Du|^2 \le 2^{-na}$$

<u>then we have</u>

$$\int_{B(x,2^{-n}) \times ([t-2^{-2n},t] \cap R^+)} |u||Du| \le C2^{(-3-\varepsilon)n} \quad \text{if } 0 < t \le T.$$

Proof. We set $b = (-2an - 5pn + 10n)/p$. Using the inequality $cd \le (1/2)ec^2 + (1/2)e^{-1}d^2$, Hölder's inequality, the hypotheses, and the definitions of a and b, we set $E = B(x,2^{-n}) \times ([t-2^{-2n},t] \cap R^+)$ and compute

$$\int_E |u||Du| \le (1/2)\int_E 2^{(-na-b)/2}|u|^2 + (1/2)\int_E 2^{(na+b)/2}|Du|^2$$

$$\le (1/2)2^{(-na-b)/2}\left(\int_E (|u|^2)^{p/2}\right)^{2/p}\left(\int_E 1^{p/(p-2)}\right)^{(p-2)/p}$$

$$+ (1/2)2^{(na+b)/2}\int_{B(x,2^{-n}) \times [0,T]} |Du|^2$$

$$\le C2^{(-na-b)/2}(2^{-na})^{2/p}(2^{-5n})^{(p-2)/p}$$

$$+ (1/2)2^{(na+b)/2}2^{-na}$$

$$= C2^{(-3-\varepsilon)n} + (1/2)2^{(-3-\varepsilon)n} = C2^{(-3-\varepsilon)n}.$$

§5. Estimates using Hausdorff measure.

Definition 5.1. For $0 < a < \infty$ and $2 < p < 10/3$, let $A(a,p)$ be the collection of all points $x \in R^3$ for which there exists a positive integer N (depending on x) such that

$$\int_{B(x,2^{-n}) \times [0,T]} |u|^p \le 2^{-an} \quad \text{and}$$

$$\int_{B(x,2^{-n}) \times [0,T]} |Du|^2 \le 2^{-an}$$

hold whenever n is an integer with $n \ge N$.

Lemma 5.1. *The a dimensional Hausdorff measure of $R^3 - A(a,p)$ is finite.*

Proof. Let $\varepsilon > 0$ be given. For each $x \in R^3 - A(a,p)$ there exists a ball $B_x = B(x,r_x)$, where $r_x \le \varepsilon$, such that the inequality

(5.1) $\int_{B_x \times [0,T]} |u|^p + \int_{B_x \times [0,T]} |Du|^2 \ge (r_x)^a$

holds. Let $Z = \{B_x : x \in R^3 - A(a,p)\}$. We use Lemma 2.1 to find Z_i satisfying parts 1), 2), and 3) of that lemma. For each i we use (5.1) and parts 2), 1) of Lemma 2.1 to compute

(5.2) $\int_{R^3 \times [0,T]} |u|^p + \int_{R^3 \times [0,T]} |Du|^2$

$$\ge \Sigma_{B \in Z_i} \int_{B \times [0,T]} |u|^p + \Sigma_{B \in Z_i} \int_{B \times [0,T]} |Du|^2$$

$$\ge \Sigma_{B \in Z_i} (\text{radius}(B))^a = 2^{-a} \Sigma_{B \in Z_i} (\text{diameter}(B))^a.$$

Summing (5.2) over i = 1, 2, ... , K and using (1.1) and Lemma 4.3

(recall that A(a,p) is defined only for $2 < p < 10/3$) we obtain

(5.3) $$\sum_{B \in \bigcup_{i=1}^{K} Z_i} (\text{diameter}(B))^a \leq 2^a KP$$

where P is a finite constant that does not depend on ε. In addition, part 3) of Lemma 2.1 and the definition of Z imply

(5.4) $$R^3 - A(a,p) \subset \bigcup_{i=1}^{K} Z_i.$$

Combining (5.3), (5.4), and the fact that the diameters of the balls in Z do not exceed 2ε, we obtain $\phi^a_{2\varepsilon}(R^3 - A(a,p)) \leq 2^a CP$ (see §2). Since $2^a CP$ does not depend on ε, we conclude from §2 that $\mathcal{H}^a(R^3 - A(a,p))$ is finite.

§6. An inequality for the function u.

We define functions ϕ and ψ with domain $R^3 \times \{t: t < 0\}$ and range R^+ as follows:

$$\phi(x,t) = (2\sqrt{\pi})^{-3}(-t)^{-3/2}\exp(|x|^2/(4t))$$

$$\psi(x,t) = -(4\pi)^{-1}\int_{R^3} \phi(y,t)\,|y - x|^{-1}\,dy.$$

These definitions were used in [S2]. We have the following result:

<u>Lemma 6.1.</u> <u>Let</u> $t \in \bigcup_{q\geq 0} J_q$ <u>be given</u>. <u>Then for</u> $i \in \{1,2,3\}$ <u>and</u> $x \in R^3$ <u>the following identity holds</u>:

$$u_i(x,t) = \int_{R^3} u_i(y,0)\phi(y - x,-t)\,dy$$
$$- \int_0^t \int_{R^3} u_j(y,s)u_{i,j}(y,s)\phi(y - x,s - t)\,dy\,ds$$
$$+ \int_0^t \int_{R^3} u_j(y,s)u_{k,j}(y,s)\psi_{,ik}(y - x,s - t)\,dy\,ds.$$

Proof. A trivial modification of the proof of Lemma 1.1 of [S2] yields

$$u_i(x,t) = \int_{R^3} u_i(y,0)\phi(y - x,-t)\,dy$$
$$+ \int_0^t \int_{R^3} u_j(y,s)u_i(y,s)\phi_{,j}(y - x,s - t)\,dy\,ds$$
$$- \int_0^t \int_{R^3} u_j(y,s)u_k(y,s)\psi_{,ijk}(y - x,s - t)\,dy\,ds.$$

Now integration by parts, (1.1), and (1.2) yield the conclusion of Lemma 6.1.

<u>Lemma 6.2.</u> <u>Let</u> $t \in \bigcup_{q\geq 0} J_q$ <u>be given with</u> $t \leq T$. <u>If</u> $x \in R^3$ <u>and</u> N <u>is a positive integer then we have</u>

$$|u(x,t)| \leq \left|\int_{R^3} u(y,0)\phi(y-x,-t)\,dy\right| + C2^{3N}MT^{1/2}$$
$$+ C\,\Sigma_{n=N}^{\infty} 2^{3n}\int_{B(x,2^{-n})\times([t-2^{-2n},t]\cap R^+)} |u|\,|Du|.$$

Proof. It is easy to check that ϕ and ψ satisfy the estimates

$$|\phi(y,s)| \leq C(|y|^2 - s)^{-3/2} \quad\text{and}\quad |\psi_{,ik}(y,s)| \leq C(|y|^2 - s)^{-3/2}$$

(compare (1.5) of [S2]). Hence we can use $t \leq T$ to obtain

$$(6.1)\quad \left|\int_0^t \int_{R^3} u_j(y,s)u_{i,j}(y,s)\phi(y-x,s-t)\,dy\,ds\right|$$
$$+ \left|\int_0^t \int_{R^3} u_j(y,s)u_{k,j}(y,s)\psi_{,ik}(y-x,s-t)\,dy\,ds\right|$$
$$\leq C2^{3N}\int_{(R^3\times[0,t])-(B(x,2^{-N})\times([t-2^{-2N},t]\cap R^+))} |u|\,|Du|$$
$$+ C\,\Sigma_{n=N}^{\infty} 2^{3n}\int_{\left\{\begin{array}{l}(B(x,2^{-n})\times([t-2^{-2n},t]\cap R^+))\\ \quad -(B(x,2^{-(n+1)})\times([t-2^{-2(n+1)},t]\cap R^+))\end{array}\right\}} |u|\,|Du|$$
$$\leq C2^{3N}\int_{R^3\times[0,T]} |u|\,|Du|$$
$$+ C\,\Sigma_{n=N}^{\infty} 2^{3n}\int_{B(x,2^{-n})\times([t-2^{-2n},t]\cap R^+)} |u|\,|Du|.$$

In addition, the Schwarz inequality and (1.1) yield

$$(6.2)\quad \int_{R^3\times[0,T]} |u|\,|Du| \leq \left(\int_{R^3\times[0,T]} |u|^2\right)^{1/2}\left(\int_{R^3\times[0,T]} |Du|^2\right)^{1/2}$$
$$\leq (MT)^{1/2}M^{1/2}.$$

Now (6.1), (6.2), and Lemma 6.1 yield the conclusion of Lemma 6.2.

§7. Conclusion of the proof of Theorem 1.

Let $2 < p < 10/3$ and $\varepsilon > 0$ be given, and set $a = (p+10+2p\varepsilon)/(p+2)$. We take $x \in A(a,p)$ (Definition 5.1) and we let N be a positive integer corresponding to x as in Definition 5.1. If $t \in \bigcup_{q\geq 0} J_q \cap [0,T]$ is given, then Lemma 6.2, Definition 5.1, and Lemma 4.4 yield

$$(7.1)\quad |u(x,t)| \leq \left|\int_{R^3} u(y,0)\phi(y-x,-t)\,dy\right| + C2^{3N}MT^{1/2} + C\,\Sigma_{n=N}^{\infty} 2^{-\varepsilon n}.$$

The first term on the right side of (7.1) is bounded for $t \in [0,T]$ because $\phi(y,-t)$ is a smoothing of a Dirac Delta function and $u(y,0)$ is assumed to be a smooth function of y. The third term is

$$C\ \Sigma_{n=N}^{\infty}\ 2^{-\varepsilon n} = C(2^{-\varepsilon N})(1 - 2^{-\varepsilon})^{-1}.$$

Therefore $|u(x,t)|$ is a bounded function of $t \in \bigcup_{q\geq 0} J_q \cap [0,T]$. This proves $A(a,p) \subset A$ (see Theorem 1), and hence (see Lemma 5.1) $\mathcal{H}^a(R^3 - A) \leq \mathcal{H}^a(R^3 - A(a,p)) < \infty$. The number a can be made arbitrarily close to 5/2 by choosing p sufficiently close to 10/3 and making ε sufficiently small. Therefore the definition in §2 implies that the Hausdorff dimension of $R^3 - A$ is at most 5/2. This proves Theorem 1.

It can be shown that Theorem 1 still holds if the definition of A is replaced by $A = \{x \in R^3$: the restriction of u to $\{x\} \times [0,T]$ is equal almost everywhere (Lebesgue 1-dimensional measure) to a continuous function$\}$. The proof is a more complex version of the proof that has been presented here.

References

[F] H. Federer, Geometric Measure Theory, Springer-Verlag, New York, 1969.

[LJ] J. Leray, Sur le mouvement d'un liquide visqueux emplissant l'espace, Acta Math. 63 (1934), 193-248.

[LO] O. A. Ladyzhenskaya, The Mathematical Theory of Viscous Incompressible Flow, revised English edition, Gordon & Breach, New York, 1964.

[M] B. Mandelbrot, Les Objets Fractals, Flammarion, Paris, 1975.

[S1] V. Scheffer, Geometrie fractale de la turbulence. Equations de Navier-Stokes et dimension de Hausdorff, C. R. Acad. Sc. Paris, 282 (Jan. 12, 1976), Serie A - 121-122.

[S2] __________, Partial regularity of solutions to the Navier-Stokes equations, to appear in the Pacific Journal of Mathematics.

LOCAL EXISTENCE OF $\mathscr{C}^\infty$ SOLUTIONS OF THE EULER EQUATIONS OF INCOMPRESSIBLE PERFECT FLUIDS

R. TEMAM

Département Mathématiques, Université de Paris-Sud

91405 - Orsay, France

Let Ω be an open set in $\mathbb{R}^n$ (any $n\geq 2$) , and let $\Gamma = \partial\Omega$ denotes its boundary. We consider the Euler equations for a fluid filling Ω :

$$\frac{\partial u}{\partial t} + \sum_{i=1}^{n} u_i \frac{\partial u}{\partial x_i} + \text{grad } p = f \quad \text{in} \quad \Omega \times (0,T) \tag{0.1}$$

$$\text{div } u = 0 \quad \text{in} \quad \Omega \times (0,T) \tag{0.2}$$

$$u.\nu = 0 \quad \text{on} \quad \Gamma \times (0,T) \tag{0.3}$$

$$u(x,0) = u_o(x) \quad \text{in} \quad \Omega , \tag{0.4}$$

where f and u_o are given, $u = (u_1,\dots,u_n)$ and p are unknown functions on $\Omega \times (0,T)$.

The set Ω will be either the whole space $\mathbb{R}^n$, or a bounded set with a suf-ficently regular boundary ; with some minor modifications, Ω may be as well the complement of a regular bounded set.

We show here the existence and uniqueness of $\mathscr{C}^\infty$ solutions of (0.1)-(0.4) defined on an arbitrary interval of time if $n = 2$, on a short interval of time if $n\geq 3$. We repeat with some slight modifications the proof given in [5]-[12] ; the proof is based on an a priori estimate. A totaly different proof of this result was given in [4]. Using Schauder estimates instead of Sobolev estimates one can obtain with the method used here, weaker results of regularity i.e. existence of solutions in spaces of Hölder continuous functions ; see also [2].

1. Some functional inequalities.

1.1. The inequalities.

Let $H^s(\mathbb{R}^n)$ be the real Sobolev space of order s , $s \in \mathbb{R}$, $s\geq 0$, i.e.

$$\{u \in L^2(\mathbb{R}^n) \ , \ |\xi|^{s/2} \hat{u} \in L^2(\mathbb{R}^n_\xi)\} \tag{1.1}$$

which is a Hilbert space for the norm

$$\|u\|_s = \{\int(1+|\xi|^{2s})|\hat{u}(\xi)|^2\, d\xi\}^{1/2}, \tag{1.2}$$

where $\hat{u} = \mathcal{F}u$ denotes the Fourier transform of u.

For $s = 0$, $H^o(\mathbb{R}^n) = L^2(\mathbb{R}^n)$ and we write

$$\|u\|_o = |u| . \tag{1.3}$$

We denote by A the operator $(2\pi)^{-1}(-\Delta)^{1/2}$:

$$Au = \mathcal{F}^{-1}(|\xi|\,\mathcal{F}u) . \tag{1.4}$$

Its sth power, A^s, is defined by

$$A^s u = \mathcal{F}^{-1}(|\xi|^s \mathcal{F}u) , \ \forall\, u \in H^s(\mathbb{R}^n) . \tag{1.5}$$

We set

$$[u]_s = |A^s u| , \tag{1.6}$$

and then

$$\|u\|_s = ([u]_s^2 + |u|^2)^{1/2} , \quad \forall\, s>0 . \tag{1.7}$$

LEMMA 1.1. *We assume that* u, v *belong to* $H^s(\mathbb{R}^n)$, $s>1+n/2$, *and let* γ *be any number,* $\frac{n}{2} < \gamma \leq s-1$, *then*

$$|A^s(uv)-uA^s v| \leq c_o(s,\gamma,n)\, \{\|u\|_s\|v\|_\gamma + \|u\|_{\gamma+1}\|v\|_{s-1}\} , \tag{1.8}$$

where $c_o = c_o(s,\gamma,n)$ *depends only on* s, γ, n .

Remark 1.1. i) In the proof of Lemma 1.1 given below we will see the two following improved form of (1.8) : under the same assumptions,

$$|A^s(uv)-uA^s v| \leq s\, \max(2^{s-2},1)\, \{[u]_s\|v\|_{\mathcal{F}^{-1}L^1} + \|Du\|_{\mathcal{F}^{-1}L^1}\, [v]_{s-1}\} . \tag{1.9}$$

$$|A^s(uv)-uA^s v| \leq c_1(s,\gamma,n)\, \{[u]_s\|v\|_\gamma + \|u\|_{\gamma+1}[v]_{s-1}\} , \tag{1.10}$$

where c_1 depends only on γ and n , and

$$\|\phi\|_{\mathcal{F}^{-1}L^1} = \|\hat{\phi}\|_{L^1(\mathbb{R}^n)} .$$

ii) An inequality similar to (1.10) and (1.8) holds if we replace A^s by any differential operator $D^\alpha = D_1^{\alpha_1} \ldots D_n^{\alpha_n}$, of order $s = \alpha_1 + \ldots + \alpha_n$ ($D_i = \partial/\partial x_i$) . ∎

Now let Ω be a bounded domain in $\mathbb{R}^n$ whose boundary Γ is of class $\mathcal{C}^r$, r sufficently large.

The Sobolev space $H^s(\Omega)$, $s \in \mathbb{R}$, $s \geq 0$, is defined either by interpolation as in Lions-Magenes [9], or as the space of restrictions to Ω of the functions v in $H^s(\mathbb{R}^n)$. We endow $H^s(\Omega)$ with the Hilbert norm

$$(1.11) \qquad \|u\|_s = \inf \{\|v\|_{H^s(\mathbb{R}^n)} , v \in H^s(\mathbb{R}^n), v|_\Omega = u\} .$$

For $s = m$ integer, this norm is equivalent to the norm

$$(|u|^2 + [u]_m^2)^{1/2}$$

where

$$(1.12) \qquad [u]_m^2 = \sum_{|\alpha|=m} |D^\alpha u|^2$$

($|.|$ = norm in $L^2(\Omega)$) .

The following versions of a well-known inequality will be useful.

LEMMA 1.2. *For $s > 1+n/2$, $H^s(\Omega)$ is an algebra for the pointwise multiplication and*

$$(1.13) \qquad \|uv\|_s \leq c_2(s,\gamma,\Omega) \{\|u\|_\gamma \|v\|_s + \|u\|_s \|v\|_\gamma + \|u\|_{\gamma+1} \|v\|_{s-1}\} ,$$

$\forall u, v \in H^s(\Omega)$, and for an arbitrary γ , $\frac{n}{2} < \gamma \leq s-1$.

Finally

LEMMA 1.3. *We assume that u, v belong to $H^s(\Omega)$ s integer, $s > 1+n/2$ and let γ be any number, $\frac{n}{2} < \gamma \leq s-1$.*

Then $\forall u, v \in H^s(\Omega)$,

$$(1.14) \qquad |D^\alpha(uv) - uD^\alpha v| \leq c_3(s,\gamma,\Omega) \{\|u\|_s \|v\|_\gamma + \|u\|_{\gamma+1} \|v\|_{s-1}\}$$

where $D^\alpha = D_1^{\alpha_1} \ldots D_n^{\alpha_n}$, with $\alpha_1 + \ldots + \alpha_n = s$.

1.2. The proofs.

The reading of the proofs is not necessary for the sequel.

Proof of Lemma 1.1. (cf.[11]). We have

$$\widehat{A^s(u.v)}(\xi) = |\xi|^s \, \widehat{uv}(\xi) = \int |\xi|^s \, \hat{u}(\xi-\xi') \, \hat{v}(\xi') \, d\xi'$$

and

$$\widehat{u\, A^s v}(\xi) = \int \hat{u}(\xi-\xi') |\xi'|^s \, \hat{v}(\xi') \, d\xi' \; .$$

The left hand side of (1.8) is the L^2 norm of $Y = A^s(uv) - u A^s v$ or, using Parseval formula, the L^2 norm of $\hat{Y}$:

$$\hat{Y}(\xi) = \int (|\xi|^s - |\xi'|^s) \, \hat{u}(\xi-\xi') \, \hat{v}(\xi') \, d\xi' \; .$$

It is easy to check that

$$\begin{aligned} ||\xi|^s - |\xi'|^s| &= \int_0^1 \frac{d}{d\lambda} (|\lambda\xi+(1-\lambda)\xi'|^s) \, d\lambda \\ &= s \int_0^1 |\lambda\xi+(1-\lambda)\xi'|^{s-2} \, (\lambda\xi+(1-\lambda)\xi')(\xi-\xi') \, d\lambda \\ &\leqslant s(|\xi|+|\xi'|)^{s-1}|\xi-\xi'| \\ &\leqslant s.\max(1,2^{s-2}) \, (|\xi|^{s-1} + |\xi'|^{s-1})|\xi-\xi'| \; . \end{aligned}$$

Whence

$$|\hat{Y}(\xi)| \leqslant s.\max(2^{s-2},1)(Y_1(\xi) + Y_2(\xi))$$

$$|Y| = |\hat{Y}| \leqslant s.\max(2^{s-2},1)(|Y_1|+|Y_2|) \; , \tag{1.15}$$

where

$$Y_1(\xi) = \int |\xi-\xi'|^s \, |\hat{u}(\xi-\xi')| \, |\hat{v}(\xi')| \, d\xi' \; ,$$

$$Y_2(\xi) = \int |\xi-\xi'|^s \, |\xi'|^{s-1} \, |\hat{u}(\xi-\xi')| \, |\hat{v}(\xi')| \, d\xi' \; .$$

The functions Y_1 , Y_2 are both convolution products and, using the convolution inequalities we find

$$|Y_1| \leqslant ||\xi|^s|\hat{u}(\xi)|| \, \|\hat{v}\|_{L^1(R_\xi)}$$

(1.16) $$|Y_1| \leq [u]_s \, \|\hat{v}\|_{L^1(\mathbb{R}_\xi)} ,$$

$$|Y_2| \leq \| |\xi|^{s-1} |\hat{v}(\xi)| \| \, \| |\xi| |\hat{u}(\xi)| \|_{L^1(\mathbb{R}_\xi)}$$

(1.17) $$|Y_2| \leq [v]_{s-1} \|\widehat{Du}\|_{L^1(\mathbb{R}_\xi)} .$$

Whence (1.9). The passage of (1.9) to (1.10) and (1.8) corresponds to the majoration of the norm $\|\hat{w}\|_{L^1(\mathbb{R}_\xi)}$ of a function w, in term of a Sobolev norm of w. The simplest result is

$$\int |\hat{w}(\xi)| \, d\xi = \int |\hat{w}(\xi)| (1+|\xi|^{2\gamma})^{1/2} \frac{d\xi}{(1+|\xi|^{2\gamma})^{1/2}}$$

$$\leq \left(\int \frac{d\xi}{1+|\xi|^{2\gamma}}\right)^{1/2} \left(\int (1+|\xi|^{2\gamma}) |\hat{w}(\xi)|^2 \, d\xi\right)^{1/2}$$

$$\leq c_4(\gamma,n) \|w\|_\gamma ,$$

provided

$$c_4 = \left(\int \frac{d\xi}{1+|\xi|^{2\gamma}}\right)^{1/2} < +\infty ,$$

i.e. $\gamma > n/2$.

Point ii) of Remark 1.1 follows easily too ; we just replace above $|\xi|^s$ by $\xi^\alpha = \xi_1^{\alpha_1} \ldots \xi_n^{\alpha_n}$.

Proof of Lemma 1.2. There exists a prolongation operator P, linear and continuous from $H^s(\Omega)$ into $H^s(\mathbb{R}^n)$, its left inverse R, the restriction operator being continuous from $H^s(\mathbb{R}^n)$ into $H^s(\Omega)$ (cf. [9]).

If $u, v \in H^s(\mathbb{R}^n)$, then uv is the restriction to Ω of $Pu.Pv$, and

$$\|uv\|_{H^s(\Omega)} = \|R(Pu.Pv)\|_{H^s(\Omega)}$$

$$\leq c_1' \|Pu.Pv\|_{H^s(\mathbb{R}^n)}$$

$$= c_2' (|Pu.Pv|^2 + [Pu.Pv]_s^2)^{1/2} .$$

Now $H^\gamma(\mathbb{R}^n)$ is included in the space of continuous bounded functions on $\mathbb{R}^n$ and

$$|Pu.Pv| \leq c_3' \|Pu\|_\gamma |Pv|$$

$$\leq c_4' \|u\|_\gamma |v| .$$

Using Lemma 1.1 we also have

$$[Pu.Pv]_s = |A^s(Pu.Pv)| \leqslant |Pu.A^s(Pv)| + |A^s(Pu.Pv)-Pu.A^s(Pv)|$$

$$\leqslant \|Pu\|_\gamma \|Pv\|_s + c_o\{\|Pu\|_s\|Pv\|_\gamma + \|Pu\|_{\gamma+1}\|Pv\|_{s-1}\}$$

$$\leqslant c_5'\{\|u\|_\gamma\|v\|_s + \|u\|_s\|v\|_\gamma + \|u\|_{\gamma+1}\|v\|_{s-1}\} .$$

Proof of Lemma 1.3. We note that

$$|D^\alpha(uv) - uD^\alpha v|_{L^2(\Omega)} \leqslant |D^\alpha(Pu.Pv) - Pu.D^\alpha(Pv)|_{L^2(\mathbb{R}^n)}$$

$\leqslant$ (by Lemma 1.1 and Remark 1.1 ii))

$$\leqslant c_1 \{\|Pu\|_s\|Pv\|_\gamma + \|Pu\|_{\gamma+1}\|Pv\|_{s-1}\}$$

$$\leqslant c_6' \{\|u\|_s\|v\|_\gamma + \|u\|_{\gamma+1}\|v\|_{s-1}\} .$$

2. Existence, Uniqueness and regularity of the solution.

2.1. The main Theorem.

THEOREM 2.1. *i) Assume that* $\Omega = \mathbb{R}^n$ *or that* Ω *is a bounded open set of* $\mathbb{R}^n$ *of class* $\mathcal{C}^{s+2}$.

For u_o *given in*

$$X_s = \{u \in H^s(\Omega)^n ,\ \nabla u = 0 ,\ u.\nu = 0 \text{ on } \Gamma\} , \tag{2.1}$$

$s > (n/2) + 1$ *and* f *given in* $L^1([0,\infty[; H^s(\Omega)^n)$ *, there exists* T_* *depending on* u_o *,* f *and* Ω *, there exist a unique*

$$u \in L^\infty(0,T_*; H^s(\Omega)^n) ,\ p \in L^\infty(0,T_*; H^{s-1}(\Omega)) ,$$

which are solutions of the Euler equations.

ii) $T_* = +\infty$ *if* $n = 2$ *and if* $n \geqslant 3$ *the number* T_* *is independant of* s *and is* $\geqslant T_1$ *, for any* T_1 *such that*

$$\int_0^{T_1} \|grad\ u(t)\|_{\mathcal{F}^{-1}L^1}\, dt < +\infty \qquad (\Omega = \mathbb{R}^n) , \tag{2.2}$$

or

$$(2.3)\qquad \int_0^{T_1} \|u(t)\|_\gamma \, dt < +\infty \quad \textit{for some} \quad \gamma > 1 + \frac{n}{2} \quad (\Omega \neq \mathbb{R}^n) .$$

iii) If Γ *is of class* $\mathscr{C}^\infty$, $u_0 \in \mathscr{C}^\infty(\bar{\Omega})^n \cap X_0$, *and* $f \in \mathscr{C}^\infty([0,\infty) \times \bar{\Omega})^n$, *then* $u \in \mathscr{C}^\infty([0,T_*] \times \bar{\Omega})^n$, $p \in \mathscr{C}^\infty([0,T_*] \times \bar{\Omega})$, *with the same* T_* *as before.*

Remark 2.2. For $\Omega \neq \mathbb{R}^n$, the point ii)-(2.2) is still valid if we set

$$\|w\|_{\mathscr{F}^{-1}L^1} = \operatorname{Inf} \{ \|\hat{\phi}\|_{L^1(\mathbb{R}^n_\xi)} , \quad \phi \in L^2(\mathbb{R}^n) , \ \phi = w \ \text{ on } \ \Omega \} .$$

2.2. A priori estimate.

Assume that u, p are sufficently regular solutions to the Euler equations. Taking the scalar product of (0.1) with u in $L^2(\Omega)^n$, we find

$$(2.4)\qquad (\tfrac{1}{2}) \frac{d}{dt} |u(t)|^2 = (f(t),u(t)) .$$

If $\Omega = \mathbb{R}^n$ we apply A^s to each side of (0.1) and take the L^2-scalar products

$$\frac{1}{2}\frac{d}{dt} [u(t)]_s^2 = [f(t),u(t)]_s - \sum_{i,j=1}^n (A^s(u_i . D_i u_j), A^s u_j)$$

$$= [f(t),u(t)]_s - \sum_{i,j=1}^n (A^s(u_i D_i u_j) - u_i D_i A^s u_j, A^s u_j)$$

since $\sum_{i=1}^n (u_i D_i v, v) = 0$. With (1.8) and (2.4) we obtain for $\gamma > 1 + \frac{n}{2}$:

$$\frac{1}{2}\frac{d}{dt} \|u(t)\|_s^2 \leq c_1' \|u(t)\|_\gamma \|u(t)\|_s^2 + \|f(t)\|_s \|u(t)\|_s$$

or after simplification

$$(2.5)\qquad \frac{d}{dt} \|u(t)\|_s \leq c_1' \|u(t)\|_\gamma \|u(t)\|_s + \|f(t)\|_s .$$

If $\Omega \neq \mathbb{R}^n$, we apply D^α to each side of (0.1), and sum for $|\alpha| = \alpha_1 + \ldots + \alpha_n = s$ Using (1.14) instead of (1.8) we obtain :

$$(2.6)\qquad \frac{1}{2}\frac{d}{dt} \|u(t)\|_s^2 \leq c_2' \|u(t)\|_\gamma \|u(t)\|_s^2 + \|\nabla p(t)\|_s \|u(t)\|_s + \|f(t)\|_s \|u(t)\|_s .$$

In order to estimate $\|\nabla p(t)\|_s \leq \|p(t)\|_{s+1}$, we use the representation of p in term of u (cf. [2] [3]) :

(2.7)
$$\begin{cases} \Delta p = \operatorname{div} f - \sum_{i,j=1}^{n} D_i u_j . D_j u_i \\ \dfrac{\partial p}{\partial \nu} = f.\nu + \sum_{i,j=1}^{n} u_i u_j \ \phi_{ij} \end{cases}$$

where $\phi_{ij}(x) = \dfrac{D_{ij}\ \phi(x)}{|\operatorname{grad}\ \phi(x)|}$ depends only on Γ ($\phi(x) = 0$ represents locally the equation of Γ). Now by [1],

$$\|p\|_{s+1} \leq c(s,\Omega)\ \{\|\Delta p\|_{s-1} + |\tfrac{\partial p}{\partial \nu}|_{H^{s-1/2}(\Gamma)}\}$$

$\leq$ (by (2.6) and the trace theorem)

(2.8)
$$\leq c(s,\Omega)\ \{\|f\|_s + \sum_{i,j=1}^{n} \|D_i u_j . D_j u_i\|_{s-1} + \sum_{i,j=1}^{n} \|u_i u_j\ \phi_{ij}\|_s\}$$

$\leq$ (by (1.13), for $\gamma > 1 + n/2$)

$$\leq c_3'\ \{\|f\|_s + \|u\|_\gamma \|u\|_s\} \quad (1)$$

From this and (2.6) we easily derive an inequality similar to (2.5).

2.3. <u>The perturbed problems</u>.

The existence proof based on this a priori estimate was obtained by T. Kato [7] by passing to the limit $\nu \longrightarrow 0$ (ν = the viscosity) in the Navier Stokes equations. This method <u>cannot</u> apply to the bounded case because of the boundary layer difficulties. We are going to show existence by a different singular perturbation method, closely related to an idea used in [2] where we used a Galerkin method, with a special basis.

Since $X_s \subset X_o$ with a dense continuous embedding, we have

$$X_s \subset X_o \subset X_s'$$

with dense continuous embeddings. By the Riesz representation theorem, there exists a linear continuous operator $B \in \mathcal{L}(X_s, X_s')$, such that

(2.9)
$$((u,v))_s = (Bu,v)\ , \ \forall u,\ v \in X_s\ .$$

(1) (1.13) is interesting for "large" s. For $1 + \frac{n}{2} < \gamma \leq s \leq \gamma+1$, a slightly improved form of (1.13) is necessar.

Replacing if necessar $\|\cdot\|_s$ by an equivalent norm, we may assume that

$$Bu = (-\Delta)^s u + u \,, \quad \forall\, u \text{ with compact support .} \tag{2.10}$$

For u_0, f given as in Theorem 2.1, and for an arbitrary $\varepsilon>0$ fixed, it is easy to see (cf. Lions [10]) that there exists a unique u_ε

$$u_\varepsilon \in L^2(0,T;\, X_s) \cap L^\infty(0,T;\, X_0) \tag{2.11}$$

such that

$$u_\varepsilon' + \varepsilon\, Bu_\varepsilon + P[(u_\varepsilon,\nabla)u_\varepsilon] = Pf \,. \tag{2.12}$$

$$u_\varepsilon(0) = u_0 \,, \tag{2.13}$$

P = the orthogonal projector in $L^2(\Omega)^n$ onto X_0 .

With a classical lemma characterizing the gradients (see |8| or [12] proposition I.1.1) we may interpret the abstract equation (2.12) as follows : there exists a distribution p_ε on $\Omega \times (0,\infty)$ such that

$$\frac{\partial u_\varepsilon}{\partial t} + \varepsilon[(-\Delta)^s u_\varepsilon + u_\varepsilon] + (u_\varepsilon.\nabla)u_\varepsilon + \nabla p_\varepsilon = f \quad \text{in } \Omega \times (0,T), \tag{2.14}$$

$$\operatorname{div} u_\varepsilon = 0 \quad \text{in } \Omega \times (0,T), \tag{2.15}$$

and the boundary conditions are :

$$u_\varepsilon.\nu = 0 \quad \text{on } \Gamma, \tag{2.16}$$

$$\Delta^j u_\varepsilon = 0 \text{ on } \Gamma\,, \ \frac{s}{2} \leqslant j \leqslant s-1\,,\ s \text{ odd},\ \frac{s+1}{2} \leqslant j \leqslant s-1\,,\ s \text{ even}\,, \tag{2.17}$$

$$\frac{\partial \Delta^j u_\varepsilon}{\partial \nu} = 0 \text{ on } \Gamma\,, \ \frac{s}{2} \leqslant j \leqslant s-2,\ s \text{ odd},\ \frac{s-1}{2} \leqslant j \leqslant s-2\,,\ s \text{ even}\,, \tag{2.18}$$

$$\frac{\partial \Delta^{s-1} u_\varepsilon}{\partial \nu} = \nu\Big(\frac{\partial \Delta^{s-1} u_\varepsilon}{\partial \nu}.\nu\Big) \,. \tag{2.19}$$

The a priori estimates on u_ε are obtained by taking the scalar product of (2.14) with u_ε in X_s . We obtain as for (2.5)

(2.14)

$$\frac{1}{2}\frac{d}{dt}\|u_\varepsilon(t)\|_s^2 + \varepsilon|Bu_\varepsilon(t)|^2 \leqslant c_1'\|u_\varepsilon(t)\|_\gamma \|u_\varepsilon(t)\|_s^2 + \|f(t)\|_s \|u(t)\|_s \,. \tag{2.20}$$

Whence

$$\frac{d}{dt}\|u_\varepsilon(t)\| \leqslant c_1'\|u_\varepsilon(t)\|_s^2 + \|f(t)\|_s \ , \tag{2.21}$$

and

$$u_\varepsilon \text{ remains bounded in } L^\infty(0,T_*;\ H^s(\Omega)^n) \ , \tag{2.22}$$

for any T_* such that $\sup_{t\in[0,T_*]} y(t) < +\infty$, where

$$\begin{cases} \frac{dy}{dt}(t) = c_1' \ y(t) + \|f(t)\|_s \\ y(0) = \|u_0\|_s \ . \end{cases} \tag{2.23}$$

It is easy to see also that

$$\sqrt{\varepsilon}\ Bu_\varepsilon \text{ and } u_\varepsilon' \text{ remain bounded in } L^\infty(0,T_*;X_s') \ . \tag{2.24}$$

Then the passage to the limit $\varepsilon \longrightarrow 0$, is standard (cf. [10]-[13]) and we obtain (0.1)-(0.4) at the limit.

2.4. Proof of Theorem 2.1.

The existence in point i) has just been proved. The uniqueness is standard. The point ii) for $n\geqslant 3$ follows from the estimate (2.5) and Gronwall Lemma. For $n = 2$ we use the existence result established in [6] which implies that the integral in (2.3) is bounded for any $T_1 > 0$, for some $\gamma > 1 + \frac{n}{2}$.

For point iii) we observe that $u \in L^\infty([0,T_*];H^m(\Omega)^n)$, $\forall\ m > 1 + \frac{n}{2}$, where T^* is independant of m . The $\mathscr{C}^\infty$-regularity in x follows. For the regularity in t , we differentiate (0.1) ℓ times with respect to t , and we take the scalar product with $D_t^\ell u$ in $\mathbb{H}^s(\Omega)^n$. We obtain by induction on ℓ that $\frac{\partial^\ell u}{\partial t^\ell} \in L^\infty([0,T_*];H^s(\Omega)^n)$, $\forall\ \ell\geqslant 0$, $\forall\ s > 1 + \frac{n}{2}$.

REFERENCES

[1] S. Agmon, A. Douglis, L. Nirenberg -
Estimates near the boundary for solutions of elliptic partial differential equations satisfying general boundary conditions I.
Comm. Pure, Appl. Math., vol.17, 1959, p.623-727.

[2] C. Bardos, U. Frisch - This volume.

[3] D. Ebin, J. Marsden -
Groups of diffeomorphisms and the motion of an incompressible fluid.
Ann. of Math., vol. 92, 1970, p.102-163.

[4] J. Marsden, D. Ebin, A.E. Fischer -
Diffeomorphism groups hydrodynamic and relativity.
Proceedings of the thirteenth biennal seminar of the Canadian Math. Congress, J.R. Vanstone Ed. Montreal 1972.

[5] C. Foias, U. Frisch, R. Temam -
Existence de solutions $\mathcal{C}^\infty$ *des équations d'Euler.*
C.R.A.S., t.280, série A, 1975, p.505-508.

[6] T. Kato -
On classical solutions of two dimensional nonstationary Euler equations.
Arch. Rat. Mech. Anal. vol.25, 1967, p.188-220.

[7] T. Kato -
Non stationary flows of viscous and ideal fluids in $\mathbb{R}^3$.
J. Funct. Anal., vol.9, 1972, p.296-305.

[8] O.A. Ladyzhenskaya -
The mathematical Theory of viscous incompressible flow.
Gordon and Breach, New York 1969.

[9] J.L. Lions, E. Magenes -
Non homogeneous boundary value problems and applications.
Vol.1, Springer-Verlag, Heidelberg, New York, 1972.

[10] J.L. Lions -
Quelques méthodes de résolution des prolbèmes aux limites non linéaires.
Dunod-Gauthier-Villars, Paris, 1969.

[11] J.C. Saut, R. Temam -
Remarks on the Korteweg de Vries equations.
Isr. J. Math., 24, n°1, 1976, p.78-87.

[12] R. Temam -
On the Euler equations of incompressible perfect fluids.
J. Funct. Anal., vol.20, 1975, p.32-43.

[13] R. Temam -
Navier Stokes Equations.
North Holland-Elsevier, Amsterdam-New York, 1976.